THE MOUNTAINS ARE CALLING

Environment and Region in the American West

THE MOUNTAINS ARE CALLING

Tourists and the Unmaking of Yosemite National Park

MICHAEL W. CHILDERS

University of Nebraska Press / Lincoln

Publication of this volume was assisted by Colorado State University Blake Center for the Engaged Humanities.

Acknowledgments for the use of copyrighted material appear on page 204, which constitutes an extension of the copyright page.

The University of Nebraska Press is part of a land-grant institution with campuses and programs on the past, present, and future homelands of the Pawnee, Ponca, Otoe-Missouria, Omaha, Dakota, Lakota, Kaw, Cheyenne, and Arapaho Peoples, as well as those of the relocated Ho-Chunk, Sac and Fox, and Iowa Peoples.

♾

For customers in the EU with safety/GPSR concerns, contact:
gpsr@mare-nostrum.co.uk
Mare Nostrum Group BV
Doelen 72
4831 GR Breda
The Netherlands

Library of Congress Control Number: 2026004897

Designed and set in Minion Pro by L. Welch.

Dedicated to my parents Barbara and Dale Childers

Contents

Illustrations

Maps

THE MOUNTAINS ARE CALLING

Introduction

A handful of people were milling about the Tunnel View Overlook when my wife and I pulled into the parking lot on a June evening in 2015. It had already been a long day of travel, but after setting up our camp at Crane Flats, we had decided to take a quick drive to the western edge of the Yosemite Valley before settling in for the evening. We got out of the car and walked to the overlook's edge, where we silently took in the view. The valley floor was already cast in shadow, with just the upper portion of El Capitan still ablaze in sunlight. As I stood there, soaking in as much of the scene as I could, my wife snuck behind me and began taking photographs. After shuffling back and forth, angling for a shot of the valley with no one in the foreground, she finally gave up and took several with me standing among a few small knots of people. It was a telling moment. My wife, along with every other visitor standing with us, not only failed to capture an image of the valley without people, but also failed to recognize ourselves, and our fellow visitors, as an integral part of the moment.

The valley was teeming with people the next morning when we arrived back in the Yosemite Valley. After dropping my wife off at the Vernal Falls trailhead so she could hike to the base of the falls, I drove around the valley for the next half hour seeking a parking spot. I finally secured one, and our yellow lab, Bear, and I headed out for our own little adventure. We first crossed the Sentinel Bridge over the Merced River before turning down a ponderosa-lined trail to the Curry Village. Then we crossed the Merced once again, passing by the Stoneman Meadow before finally wandering into the Yosemite Village, where we sat down and watched the crowds of tourists stream by. Banners hung throughout the village celebrating Yosemite National Park's 125th anniversary, which, in part, explained the number of people. A family passed by debating whether they wanted pizza for lunch, while an older couple stopped a ranger to ask a question. Harried tourists bustled in and out of the small grocery store grabbing what I could only imagine were last-minute items for that evening's dinner.

My wife soon returned, exhausted but smiling from her hike. The crowds had been unbelievable, she said, but seeing the falls was well worth the frustration of walking in a queue to the bottom of the Mist Trail. She went on to describe how the hike was not as she remembered it from when, when she was a child, her family would spend a week every summer in Yosemite Valley. It was often the highlight of her year, she recalled. She and her brother would escape their parents' ever-vigilant gaze to explore the valley on their bicycles and wade in the Merced River, before spending evenings sitting around a campfire. Perhaps it was a trick of memory, but she remembered Yosemite as a bit less crowded.

Others have made the same observation. In the introduction of his book, *The Tourist in Yosemite, 1855–1985*, which in part inspired this book, geographer Stanford E. Demars recalled when, in the summer of 1968, he and his wife were confronted with campgrounds so congested that visitors sat in line like vultures waiting to pounce on the next open site, the park's stores were crowded to capacity, and parking was so limited that many frustrated drivers had left their vehicles wherever they could find a spot, regardless of the signs prohibiting illegal parking. While we had reserved our campsite months in advance, the rest of the Demars's experience felt very similar nearly a half century later.[1]

A record five million people visited Yosemite National Park the summer following my wife's and my visit. Like us, most spent the majority of their time in the Yosemite Valley, whose sheer granite walls and towering waterfalls have attracted tourists since 1855. Others crossed the Sierra Nevada on the Tioga Road, perhaps stopping at Olmstead Point then driving along Lake Tenaya's northern shore, through Tuolumne Meadows, before dropping into the Yosemite Valley. That year's record number of visitors underscored both Yosemite's immense popularity and its limits. Many commented Yosemite felt more like an amusement park that summer than a crown jewel of the U.S. National Park System. Crowds inundated the park's campgrounds, viewpoints, parking lots, trailheads, and rustic village—spoiling visitors' experiences and threatening Yosemite's increasingly strained natural environment.

Following climate change, overcrowding poses the single greatest environmental threat to Yosemite. The tread of millions of feet has eroded and compacted the soil while trampling of vegetation has negatively affected water quality by hardening the surface and reducing water's ability to percolate into the ground. The sounds of cars, buses, and thousands of visitors often drown out the songs of birds. Although less severe than in urban areas, light pollution at night has dramatically impacted the behavior of nocturnal wildlife. Air pollution not only obscures the Sierra Nevada sky but also settles in

Yosemite Valley. Human-wildlife interactions have caused the deaths of tens of thousands of brown bears and changed wildlife behavior. Additionally, the park's development has intensified the effects of floods, heavy winter storms, and wildfires, increasing their destruction of the park's infrastructure. All of which have unmade Yosemite in critical ways.[2]

To understand how we got to this point, we first need to look at Yosemite's past. But rather than just cast the park as an wilderness invaded by hordes of visitors, we need to step back and realize that Yosemite's story, in many ways, is that of its visitors. I attempt to tell some of those stories in the coming pages. By placing the stories of Indigenous people, tourists, innkeepers, soldiers, rangers, climbers, concessioners, inholders, and administrators in the foreground, I seek to tell a much more complete story of the park's history. Readers will no doubt recognize many of the personalities in the coming pages, including John Muir, Frederick Law Olmsted, and Ansel Adams. But they will also meet many who have not historically been included in Yosemite's story but are very much a part of its history.

In this manner, *The Mountains Are Calling: Tourists and the Unmaking of Yosemite National Park* is a book about placemaking, and the environmental consequences of that making. Visitors have experienced Yosemite in much the same manner since the first tourists entered the Yosemite Valley in the mid-nineteenth century. Drawn by its sheer beauty, they have come to camp, hike, swim, and, most importantly, find quiet moments of reflection within the park. Through their recreation, they have not just visited but actively crafted the story of Yosemite National Park's landscape and meaning. But that meaning brought greater numbers of both visitors and development, threatening the very beauty most came to experience.[3]

Early visitors recast the Yosemite Valley, and the nearby Mariposa Grove soon afterward, as landscapes emblematic of American republican ideals, and, perhaps more crassly, as a means to attract tourists. They built lodges and roads, cut down trees, dynamited the Merced River's moraine, camped at the base of the Yosemite Falls, measured the thickness of the giant sequoia, and wrote about their experiences in both the valley and the grove in newspapers, magazines, and books. Others followed, Yosemite soon became a national park, and new roads, campgrounds, lodges, and even a ski area were built to serve growing numbers of tourists. But amid Yosemite's ever-changing landscapes, the allure of its beauty has remained a constant. Visitors, drawn by the majestic granite cliffs and the whisper of ancient trees, found solace and inspiration in its embrace, engaging in a variety of outdoor pursuits that both celebrated the park's splendor and turned Yosemite into one of the most renowned national parks.

Placemaking is never a neutral endeavor, however. The creation of Yosemite was entwined with the dispossession of its Indigenous peoples. Widely acknowledged as an act of imperialism by historians, the pushing aside of the region's Native inhabitants allowed settlers to tell a new story about the Yosemite Valley and its surrounding region, reflecting both Americans' changing views of nature and a landscape unburdened by history. This erasure of the Indigenous past allowed settlers to reinvent Yosemite as a landscape of untouched sublime wonder and contemplation. In prioritizing recreation, they not only excluded traditional practices like timber harvesting and grazing but also necessitated the preservation of the landscape's scenic values, ensuring it would be enjoyed as a park for generations to come.[4]

In this manner, Yosemite National Park has long been a symbol of American nature. From its invasion by the Mariposa Battalion in 1851 through today, its beauty and wonder have captivated generations. One runs out of superlatives in describing Yosemite, though John Muir came close to capturing its grandeur in writing that no temple made with human hands can compare to the park's wonder. Yosemite entered the American consciousness at a key moment in the nation's history. The embrace of nature's sublimity as core to American identity, along with the conquest of the American West, wove both the Yosemite Valley and the Mariposa Grove into the very fabric of our national identity. By the time it was established as a national park in 1890, Yosemite had become a renowned destination for thousands of visitors, like Muir, seeking to bask in its beauty.[5]

Yosemite is also a landscape of controversy. From legal battles over ownership and development of the Yosemite Valley, the inundation of the Hetch Hetchy Valley, and the carving of the Tioga Road out of Yosemite's granite, to fights over the park's commercialization, Yosemite's history is rife with disputes over the park's purpose, and the environmental consequences of its popularity. So while Yosemite's landscape is iconic, we cannot understand its meaning and its past without its people. From its tourists and administrators to its rangers and climbers, the park's visitors, I argue, played a central role in shaping its history. From the first party of tourists to enter the Yosemite Valley in 1855 to the millions who visit the national park today, each has driven Yosemite's popularity and, ultimately, its development. But in doing so, they came to threaten both the park's environment and the very experiences that they sought by overwhelming Yosemite's landscape, which in turn impaired its environment and visitors' ability to enjoy it.

Though no institution is more symbolic of the American conservation movement than the national parks, the environmental consequences of their popularity have increasingly threatened these hallowed places. Tasked to "promote and regulate," the National Park Service has long struggled to balance use with conservation within the parks, whose purposes are to conserve the nation's scenery, history, and wildlife unimpaired for future generations to enjoy. Yet the popularity of parks like Yosemite has challenged the Park Service's ability to meet this mandate. Overcrowding has severely affected Yosemite's natural environment, profoundly impacting visitors' enjoyment of the park. In this sense, experiencing the natural world is itself a vital natural resource, one that must be carefully managed to preserve its beauty and its accessibility.[6]

Histories of national parks are often framed as paradoxes between preservation and use, highlighting key individuals who played significant roles in the creation, preservation, or management of a particular park. These narratives, however, frequently marginalize or overlook the millions of visitors who contribute to the park's story. Most experienced Yosemite through their recreation. Yosemite became a place to escape, to hike, and to gaze at its granite cliffs and towering waterfalls. But this understanding did not emerge in a vacuum. It was influenced by shifting cultural ideas of nature, leisure, and progress. I trace how those ideas attempted to erase the park's Indigenous history, aided tourists and park managers in reshaping the landscape to meet recreational demands, and shaped debates over Yosemite's fundamental purpose.

While these traditional histories have their merits, they often neglect the broader experiences that shape the park's past. This realization led me to explore some fascinating aspects of the archives. I found a notable gap between how most visitors experience the park and environmentalists' portrayal of Yosemite as an embattled wilderness. Many families visit to enjoy a long weekend of camping, roasting marshmallows over a fire, swimming in the Merced River, and buying souvenirs at the visitor center. While they may care about the park's health, their primary concern is experiencing and enjoying the park. We often dismiss these visitors as mere crowds, failing to recognize that they represent the vast majority of those who have shaped Yosemite's history.[7]

Americans habitually place as much, if not more, value on their personal connections to places and the past as they do on scholarly works like this one. They are not mere consumers of narratives presented by others but active participants in crafting their own understandings of the past. I found much the same in exploring Yosemite's history. Visitors, Park Service personnel, and concessioners have long created their own meanings of this incomparable place, shaping history to their own wants—often while overlooking the paradox of fulfilling those desires and protecting the park's environment.[8]

While it is easy to find visitors in Yosemite, it's much more challenging to locate them in the archives. They are of course lurking in the pages of their published memoirs, letters, and newspapers. But I also found them hiding in National Park Service memoranda, reports, and planning documents. Their individual stories combine to tell Yosemite's history. Yet not all voices can be found in the archives, or in memoirs that were left behind. The archives are all too often silent concerning the experiences of countless people who played an integral role in the park's history.

Among those were the thousands of Chinese immigrants and Chinese Americans who built Yosemite's roads, washed tourists' laundry, and cooked their meals. Glimpses of their contributions can be found in brief mentions within letters, articles, and other sources, but few sources tell us their whole story. And we may never know their entire history within Yosemite. Likewise, African Americans and other minoritized groups' contributions to making the park are largely absent from the archives, even though they played as important a role as Chinese Americans in shaping Yosemite. While I attempt to correct these absences, their stories remain incomplete.[9]

This book is divided into three acts tracing Yosemite's history through its making and gradual unmaking. The first act opens with the Mariposa Battalion's invasion of Yosemite Valley in 1851. News of the remarkable valley brought a small band led by publisher and booster James Hutchings to explore the valley three years later. Within a decade, President Abraham Lincoln ceded the Yosemite Valley and Mariposa Grove to the state of California for "public use, resort, and recreation." Yet visitation to the Yosemite Grant remained sparse, with only a few hundred visitors making the arduous trip to the remote mountain valley. Chapter 2 picks up immediately following Yosemite National Park's establishment in 1890. A year later the U.S. Cavalry took control of making and managing the new national park, while growing criticisms over the despoilation of the Yosemite Valley led to the emergence of a broader struggle over the park's purpose. This story also includes the recession of the Yosemite Valley and Mariposa Grove into the national park and the battle over control of the Hetch Hetchy Valley—all of which contributed to the creation of the National Park Service in 1916.[10]

Part 2 traces the next fifty years of the park's history and the increasingly acrimonious debate over tourism's environmental costs. Chapter 3 tells the story of how a broader swath of Americans embraced national parks as places not only of respite and natural wonder but of mass entertainment. The ascendance of the automobile allowed visitors to easily travel to the park, driving increased visitation and the development of facilities like Badger Pass Ski Area and the Ahwahnee Hotel. Chapter 4 picks up Yosemite's story following World War II, as thousands poured into the park only to find its roads, campgrounds, and facilities in increasing disrepair. But as the Park Service poured millions into rebuilding Yosemite, amid growing national concerns over the environment, many began to question the consequences of continued development in the park. This led to difficult, and often contentious, debates over the Park Service's mandate and, more broadly, Yosemite's meaning. While historians have long interpreted this as a struggle over the Park Service's "dual mandate," Yosemite's millions of visitors continued to view their enjoyment as the park's fundamental purpose.

The growing gulf between those competing views plays out in the book's final act. Covering a single decade, chapter 5 explains how the 1970s proved to be the turning point in Yosemite's history. Concerns about overcrowding and development came to a head, leading to shifts not only in policy but in visitors' expectations of their experiences in the park. While most agreed that too many were visiting the park, few were willing to make any sacrifices to stem the tide of visitors. Chapter 6 tells the park's increasingly complex story from the 1980s through the opening decade of the twenty-first century. Floods and wildfires combined with continued increased visitation threatened ruin throughout the park. Yet change remained elusive as few wished to cast off Yosemite's long-held promise of enjoyment for all. Visitor numbers peaked at over five million in 2016, the National Park Service's centennial, before a global pandemic led the park to finally place a limit on visitor numbers starting in 2021.

Yosemite's history is a human story, rich with the voices of its visitors, administrators, and Indigenous inhabitants. By placing these experiences in the foreground, we can gain insight into how this remarkable place came to be, the struggles over its meaning, and how we might forge a better future for this incomparable place.

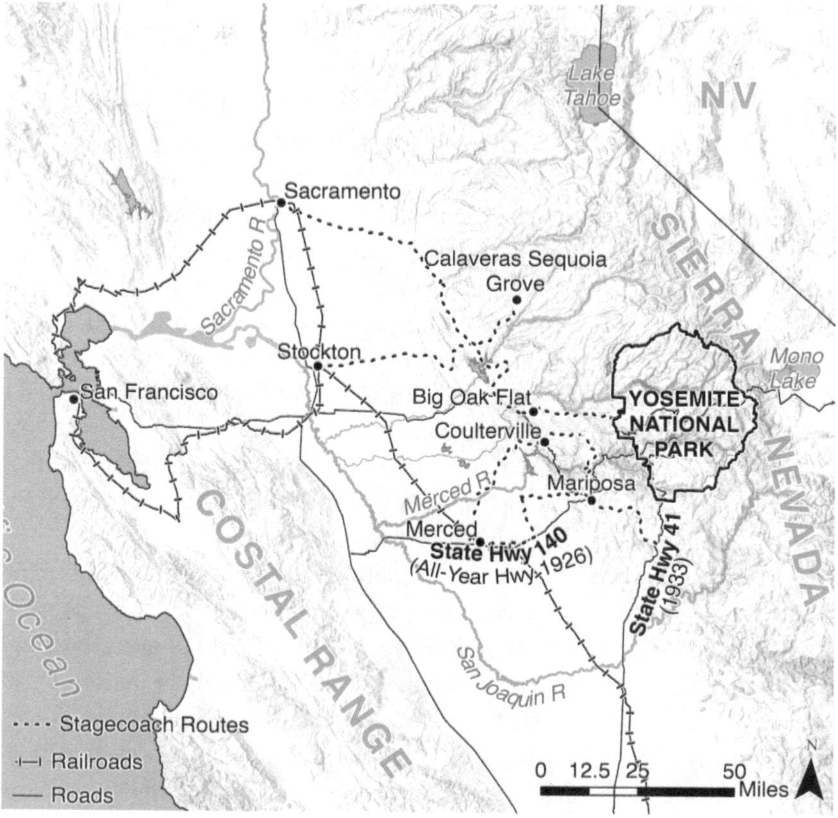

MAP 1. The San Francisco–Yosemite region. Map by Amber Bell.

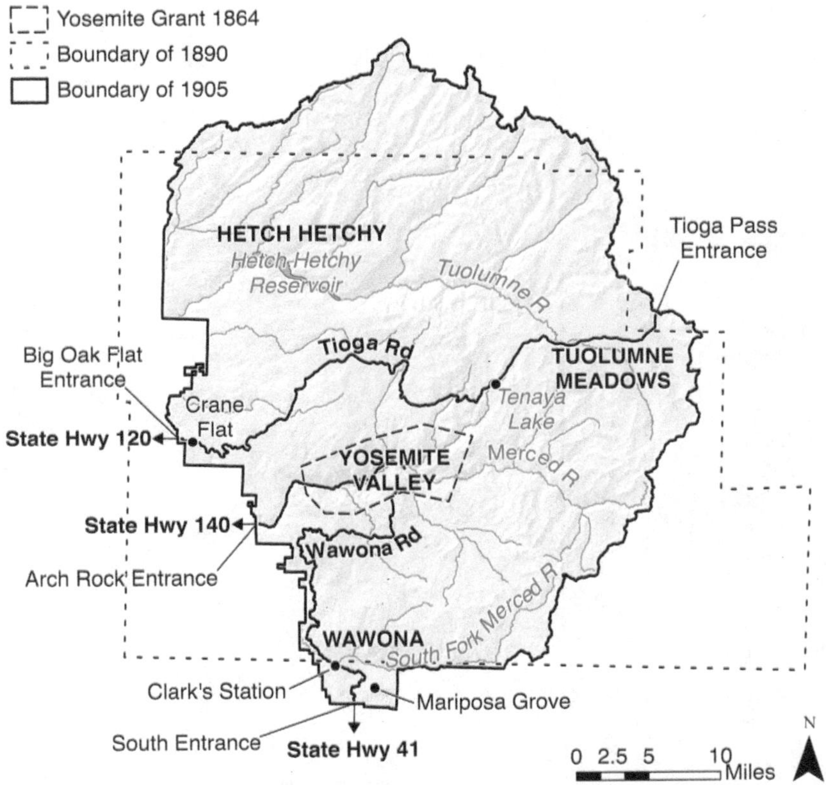

MAP 2. Yosemite National Park boundaries, 1865–1905. Map by Amber Bell.

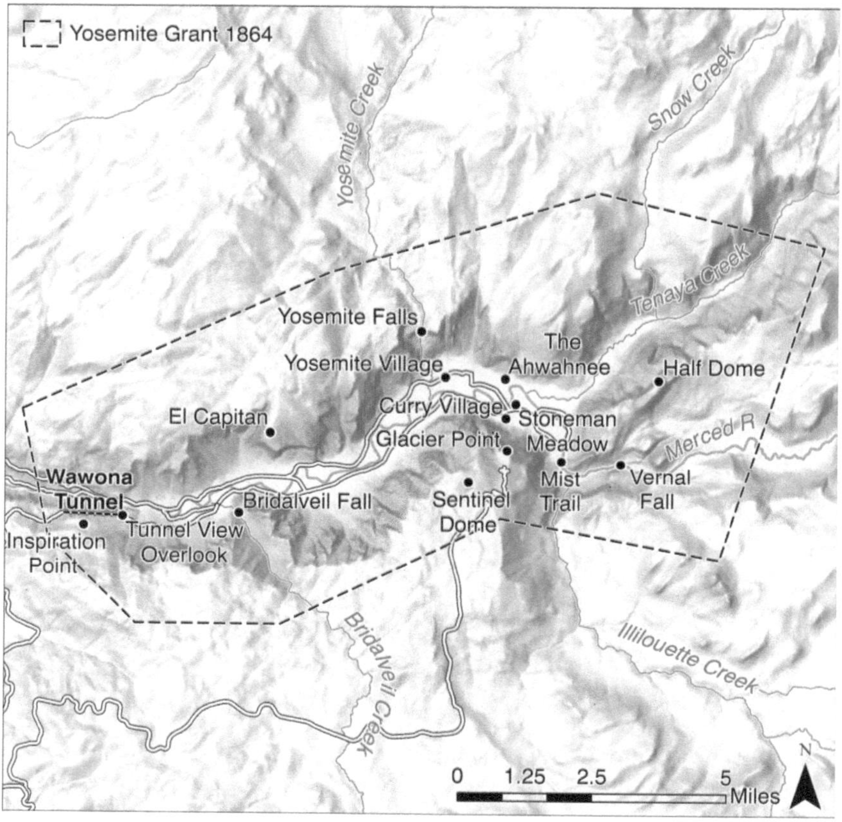

MAP 3. The Yosemite Valley. Map by Amber Bell.

PART 1

1 Making an American Landscape

A soft winter haze hung over the chasm below as wisps of clouds floated wraithlike through the trees. An immense escarpment of granite rose before Dr. Lafayette Houghton Bunnell as he sat astride his horse, spellbound by the view before him. He had left the trail above, steering his mount through the deep snow, to a rock outcropping for a clearer, and quieter, view of the valley. Awestruck by its beauty, Bunnell failed to notice the troop of cold and ill-tempered men passing him by on the trail above. A rough voice abruptly broke his silent contemplation, warning, "You'd better wake up from that dream, or you might lose your hair." Sitting at the rear of the column, Major Jim Savage, commander of the Mariposa Battalion, waited as Bunnell reluctantly turned from the view and joined the line of men slowly trudging into the valley, where they sought to capture the remaining Ahwahnechee hiding somewhere below.

The Mariposa Battalion had mustered in February 1851, on orders from California's newly elected governor, Jon McDougal, to hunt down and apprehend the small band of Miwok. War had broken out between regional Native tribes and the growing number of prospectors who had invaded the Sierra Nevada foothills in search of fortune over the past two years. Seeking retribution, a band of Miwok named the Ahwahnechee had attacked and pillaged Savage's trading post the previous fall. Enraged by the assault, local settlers demanded revenge on all Native people in the region. However, several sorties into the Sierra Nevada over the subsequent months found no sign of the band. Their leader, Tenaya, had led the Ahwahnechee to their remote valley home far in the Sierra Nevada, where the rugged mountain terrain and deep winter snows had long protected them from their enemies. But such hopes were dashed when the militia under Savage's command cornered and captured Tenaya and a small number of his band in the spring of 1851. The militia leader ordered the Ahwahnechee leader to lead the battalion to his people's refuge. Defeated, Tenaya had quietly acquiesced and led them to edge of the deep valley into which the Mariposa Battalion was now descending.[1]

Forgetting the grim task before him, Bunnell continued to marvel at the valley's beauty as he followed Savage. The major swiftly barked at him again to pay attention to the slippery, narrow path, lest he and his horse plunge off the steep hillside. Chagrined, Bunnell fell in behind Savage, and the two men joined the rest of the militia below. A light snow had covered the ground, soaking both the men and their mounts. The exhausted horses were stripped of their saddles and tack and set out to graze in a nearby meadow under the watchful eyes of a few unlucky lookouts. A campfire was soon blazing, and the remaining men huddled around it, seeking to stave off the cold. It had been an arduous day, and now all they wished for was to rest by the fire and enjoy a warm meal.

Following dinner, Bunnell wondered out loud what they should name the magnificent valley in which they were now camped. Looking up at the granite cliffs, he suggested that the valley needed a proper name—perhaps an Indian name. Others quickly joshed him for the suggestion, but then they fell quiet, taking the task seriously. After some thought, the men began offering a wide range of suggestions. Some drew from scripture, with one evoking John Milton's poem by proposing "Paradise Valley." Others offered similar names. Bunnell rejected each suggestion in turn, arguing instead that the valley needed a truly "American" name. "I could not see any necessity for going to a foreign country for a name for American scenery," Bunnell wrote when he later recounted the evening's debate, seemingly oblivious to his suggestion of an Indigenous word as the valley's "American" name.[2]

He then suggested naming the valley after the people who had lived there, the Yo-sem-i-ty. Suddenly, a hearty laugh rose above the din of voices, and before anyone else could object, the boisterous Texan John O'Neal yelled out, "Hear ye! Hear ye! Hear ye! A vote will now be taken to decide what name shall be given to this valley." After a brief show of hands, all agreed on the name Yo-sem-i-ty. And in that moment, the Ahwahnechee's home became the Yosemite Valley, at least as Bunnell later recounted of the moment.[3]

Published nearly three decades after the Mariposa Battalion's invasion of the valley, Bunnell's *Discovery of the Yosemite, and the Indian War of 1851, Which Led to That Event* joined a growing collection of travel narratives, magazine articles, and novellas documenting visitors' experiences exploring the Yosemite Valley and its surrounding region. Now living in Minnesota after having served in the Civil War, Bunnell sought to capitalize on Americans' newfound fascination with Yosemite by retelling the Mariposa Battalion's story. He readily admitted his motives in the introduction of the book, explaining that he aimed to correct existing errors and expand beyond the limits of his previous account of the battalion's entrance into the Yosemite

Valley, published in *Hutchings' California Magazine* in 1859. Bunnell's earlier version had offered a much more colorful portrayal of the militia's formation and pursuit of the Ahwahnechee into the Sierra Nevada, a tale of military conquest rather than natural wonder. This telling focused instead on the Yosemite Valley's sublimity.[4]

The Yosemite Valley had changed significantly by the time Bunnell sat down and wrote the Mariposa Battalion's story decades later. No longer an Indigenous homeland, it had become the very embodiment of American nature. It had been made more accessible by the completion of the Coulterville and Big Oak Flat Roads, as well as the transcontinental railroad, and visitation had grown from a few hundred in the mid-1860s to several thousand by the time he published *Discovery of the Yosemite, and the Indian War of 1851*.[5]

Bunnell's short history of the war and the "discovery" of the Yosemite Valley reflected American nature writing in the late nineteenth century. Greatly influenced by Ralph Waldo Emerson's 1836 essay "Nature," American writers, painters, and photographers celebrated encounters with the natural world as fundamental to American identity, culture, and even history. While such ideas initially emerged among the Northeast's middle class, they quickly extended to the American West. The region's landscapes, particularly its mountains, symbolized not only nature's grandeur but also America's republicanism—a land unique in its opportunities and beauty. Among these landscapes, none embodied this ideal more than the Yosemite Valley.[6]

Carved out of granite, the valley stretches approximately eight miles from the crest of the Sierra Nevada to its terminus below. Towering waterfalls cascade thousands of feet into the valley, feeding the Merced River; groves of ponderosa pine, black oak, and cedar scent the dry mountain air; and expansive meadows cover vast stretches of the valley's floor. For generations, the Miwok had used low-grade fires to clear the valley's underbrush, allowing them to walk easily among the trees and cultivate acorns, their primary food source. Grizzly bears once frequented the region but were driven to extinction by white hunters a century later. Brown bears, nevertheless, remain abundant today within the park, along with mule deer, hundreds of bird species, countless insects, foxes, mountain lions, and fish. Yet it is the Yosemite Valley's towering granite cliffs that have drawn visitors for more than a century.

Granite defines Yosemite, but ice created it. More than a million years ago, glaciers covered much of what is now the Sierra Nevada. These frozen rivers of ice and rock carved the broad and straight Yosemite and Hetch Hetchy valleys from the granite pluton. During the most recent glacial period, approximately 20,000 years ago, ice filled the Yosemite Valley to about a third of its depth, terminating near Bridalveil Fall. As it retreated, the glacier left behind a moraine that trapped the glacier's meltwater, creating a shallow lake. Stretching nearly six miles in length, the lake eventually filled with sediment, forming the relatively level valley floor and swampy meadows seen today. The retreating glaciers also polished the valley's granite walls, leaving behind the sheer faces that inspired Bunnell's writing.

Like countless others after him, Bunnell found it nearly impossible to capture Yosemite's scenery in words. His first attempt in 1859 barely mentioned the views or the fireside debate over the valley's naming. However, his second draft focused almost entirely on the Yosemite Valley's beauty, obscuring the militia's role in the Mariposa War. Instead, Bunnell offered a pale retelling of Emerson's *Nature*. The most read and influential essay of the nineteenth century, *Nature* argued that America's "new lands, new men, and new thoughts" distinguished it from aristocratic Europe. Emerson asserted that the nation's landscapes defined not only its people but also their very republican nature. His ideas inspired countless others, like Bunnell, in crafting a new national identity where romantic notions of nature were central to America's burgeoning middle-class ethos.[7]

Starting in the 1830s, the United States underwent a significant economic and social transformation. Driven by increasing industrialization and the institution of slavery, the country's growing wealth gave rise to an emergent middle class. Though somewhat difficult to define precisely, this new social stratum included professionals who were often critical of Federalist politics and exhibited distinctive consumerist and moralistic sentiments. Being part of the nascent American middle class involved several aspects. First, it meant having enough wealth to indulge in luxuries such as art, travel, and fashion. Influenced by the religious fervor of the nation's Second Great Awakening, many in the middle class, particularly in the Northeast, embraced social reforms, including the temperance and abolitionist movements. The availability of cheap print media, such as newspapers, magazines, and books, hastily disseminated these values among middle-class readers. Finally, the American middle class adopted an aesthetic view of nature that profoundly shaped the nation's broader cultural understanding of the natural world. Yet such narratives effectively erased others from the American landscape.[8]

A SECOND COMPANY of militia arrived in the Yosemite Valley in the spring of 1852, this time under the command of Captain John Boling. Surprising the Ahwahnechee, the troop captured five men, including one of Tenaya's sons. After a brief confrontation, Boling agreed to release a lone captive to inform Tenaya of the government's demand that the Ahwahnechee surrender and relocate to a reservation on the western edge of the San Joaquin Valley. But shortly after the messenger set out, the remaining prisoners broke free from their bonds. In the ensuing scuffle, all four men were shot and killed, including Tenaya's son. When Tenaya arrived to find his son's body lying on the ground he hastily turned to flee, but soldiers immediately seized the aging leader. Hearing the fighting, the remaining Ahwahnechee fled eastward out of the valley. The militia quickly gave chase, overtaking the small band on the shores of what would later become known as Lake Tenaya, and marched the band to the Fresno River Reservation.[9]

But the Ahwahnechee did not remain on the reservation for long. On their arrival, Tenaya implored the reservation agent to allow his people to return to their homeland, repeatedly citing the reservation's aridness, the lack of food, and the presence of their enemies as unbearable hardships. The agent finally relented to Tenaya's pleas, allowing the Ahwahnechee to return to the Yosemite Valley, provided they remained peaceful. Despite Tenaya's assurances, violence erupted again the following summer when a party of eight prospectors entered the valley. Infuriated by the intrusion into their home, the Ahwahnechee attacked the party, killing two. In response, the state once again dispatched army regulars to the valley to apprehend the Ahwahnechee and return them to the reservation.

On arriving in the valley, the troops captured and lynched several young men they believed had been involved in the skirmish. Terrified he was next, Tenaya led his remaining people over the Sierra to seek refuge with their Paiute neighbors along the shores of Mono Lake. The army pursued the band for several days before finally abandoning the chase and returning to Fort Miller. The Ahwahnechee stayed with the neighboring Paiute band for a season before returning to the Yosemite Valley. Tenaya's death the following summer, after being struck in the head in a quarrel over a game of hand bone, shattered the band. A few Ahwahnechee remained in Yosemite after his death, struggling to live as they had for centuries. To the growing number of visitors, the survivors were often seen as either a curiosity or a nuisance. But despite ongoing efforts to remove them, their descendants remained in the valley for generations.[10]

WORD QUICKLY SPREAD of the astonishing valley hidden high in the Sierra Nevada after the Mariposa Battalion's subjugation of the Ahwahnechee. Few ventured into the valley in the years immediately following the battalion's initial foray, rightfully fearing retaliation from the Ahwahnechee. However, with Tenaya's death and the conclusion of the Mariposa War in 1853, a growing number made their way into the valley. A small group of prospectors soon explored the banks of the Merced in a vain attempt to find gold, and the famed animal trapper James Adams, better known as "Grizzly Adams," unsuccessfully scouted the region in search of bears.[11]

But it was not the Yosemite Valley's mineral wealth nor its wildlife that captured James Mason Hutchings's attention. Rather, it was the tale of a thousand-foot waterfall, taller than the mighty Niagara Falls, that drew the journalist and California booster to the valley. British-born, Hutchings immigrated to New Orleans as a young man. But the discovery of gold in 1849 quickly drew him farther west. After arriving in San Francisco, Hutchings initially found work as a carpenter before moving to the rough-and-tumble town of Placerville to prospect for gold along the American River. Over the next several years, Hutchings made and lost two fortunes before abandoning prospecting altogether. With what money he had left, he purchased a handful of regional farms and invested in numerous mining operations, eventually settling in the mining community of Hangtown.

Named after the lynching of five men accused of theft and murder, the rowdy mining town in the western Sierra Nevada foothills had attracted thousands of gold seekers, merchants, and financiers by 1852. Hutchings became a fixture in the town's burgeoning cultural scene, attending its Protestant church every Sunday, joining the local temperance society, and participating in social events such as the annual Fireman's Ball. Like many of the town's growing number of middle-class reformers, Hutchings grew increasingly alarmed by the town's lawlessness and vulgarity. In response, he published "The Miner's Ten Commandments" in the local newspaper, in the spring of 1853, as a corrective to such pervasive offensive behavior. Providing prospectors ten directives, or commandments, to follow, including never having more than one claim, refraining from murdering others to steal their gains, and observing the Sabbath, the one-page litany of social mores struck a chord with reform-minded readers. Hutchings soon began selling lithographs of the commandments, opening his eyes to publishing's potential profitability.[12]

As California's gold fever waned, the state's politics increasingly mirrored those of the nation. The temperance movement gained traction, particularly in the northern part of the state, as a means to curb the drunken revelry and violence so prevalent since the gold rush began. Although California entered the Union as a free state, debates over abolition split the state. Northern Californians, particularly Republicans from the Northeast, generally supported ending slavery, while white southerners had primarily settled in Los Angeles and were far more sympathetic to the Democratic Party's defense of slavery. Beyond politics, California's growing middle class created an expanding market for books, pamphlets, and periodicals that reflected their cultural interests. Having grown up in the English bourgeoisie, Hutchings recognized this emerging market and, building on the success of the "Miner's Ten Commandments," moved to capitalize on the California middle class's developing enthusiasm for nature writing.[13]

Hutchings embarked on a tour of Northern California to gather information on the region's towns, rivers, and people, aiming to produce more lithographs in the summer of 1854. That summer, he partnered with publisher Anton Rosenfield to purchase San Francisco's Armory Hall building and start a company focused on publishing California ephemera.[14]

The following year, Hutchings set out with his close friend Walter Millard and artist Thomas Ayres to explore the Yosemite Valley. A draftsman by training, Ayres had turned to art after failing at gold prospecting. Yet, as he soon discovered, selling art in San Francisco was as challenging as panning for gold. Hutchings had befriended the struggling painter the previous summer, and he had immediately thought of Ayres when planning the trip to the enigmatic valley. Understanding that it would take more than words to convey Yosemite's beauty to readers, Hutchings knew Ayres's artistic skills would be invaluable. Although Ayres's paintings of sailing ships and landscapes had failed to attract many buyers, Hutchings believed his artist's eye would certainly help in promoting the group's journey. Flattered, Ayres eagerly agreed to join Hutchings.[15]

On the cool morning of July 5, 1855, the three men boarded a steamship in San Francisco. They first sailed to Sacramento, where they boarded a stagecoach for the ride across the Central Valley to the mining town of Mariposa, stopping along the way to visit the Calaveras giant sequoia grove. There, Hutchings's friend Alexander Stair joined the party. Traveling through California at the time proved exceedingly difficult. River voyages were dreadfully time-consuming, and often plagued by mosquitoes during the wetter months. Moreover, few roads existed beyond the Sierra Nevada foothills, and those that did were often rutted and dusty affairs, leaving riders jarred and covered in grime.

Despite these challenges, the four men reached the small community of Coulterville, where, with the help of Captain Boling, they hired the Miwok guides Kos-sum and So-pin to lead them to the remote valley. After several more days of rough mountain travel, Hutchings, Ayres, Millard, and Stair finally caught their first glimpse of the Yosemite Valley from the very precipice where Bunnell had stood just three years earlier. Later known as Inspiration Point, it would become the most iconic overlook of the Yosemite Valley.

As the others ventured down into the valley, drawn by the allure of adventure, Ayres settled in with pencil in hand and began sketching the landscape. With each deliberate stroke, he meticulously rendered a vision of the valley that future visitors would seek to experience. Half Dome loomed in the background, its grandeur subtly asserting itself over the scene. Ayers then framed the valley floor with the angularity of Sentinel Dome and the graceful cascade of Bridalveil Fall on its right, while El Capitan, imposing and unyielding, stood sentinel on its left. A scattering of pine trees and two serene meadows completed the scene. In the foreground, Ayres captured the landscape's essence by depicting a chaotic amalgamation of trees, rocks, and even the figures of his companions. In doing so, he provided a crucial perspective on the Yosemite Valley's sheer magnitude and grandeur, as well as one of the earliest portrayals of tourists within its confines.

The group spent the next five days exploring the valley. They stood at the base of El Capitan gazing up at its wide face. Across the valley, a magnificent waterfall, like a "broad, long feather of silver," flashed rainbow hues under the setting sun. "The red would mix with the purple, and the purple with the yellow, and the yellow with the green, and the green with the silvery sheen of its whitened foam, as it danced in space," Hutchings later wrote of its beauty. As the men traveled farther down the valley, they were dumbstruck by another towering waterfall, which Hutchings estimated to be at least 2,200 feet in height. They chanced upon yet another waterfall even farther down the valley, which Hutchings proclaimed to be no less than 1,500 feet tall.[16]

Ayers captured it all. His chalk and pencil drawing of Yosemite Falls depicted the falls in an almost dreamlike manner. The image showed a thin sheen of water cascading down the cliff face into a second and a third pool before finally plummeting to the valley floor. A scattering of trees provided context for the fall's height while, in the foreground, the party sat around a campfire, contemplating the valley's grandeur. Upon the men's return to San Francisco, Hutchings immediately made a lithograph of the drawing, and he featured it in the first issue of *Hutchings' California Magazine* a year later.[17]

Hutchings also penned a brief account of the trip for the newspaper the *Mariposa Gazette* upon the party's return. Then, a month later, he published a more detailed account in the *San Francisco Daily California Chronicle*. While noting that there was enough good land for several farms, Hutchings asserted the valley was not ideally suited for agriculture. Rather, he suggested it was better suited for tourists' enjoyment. A year later, Hutchings published the first edition of his magazine. Building on his success with lithographs, *Hutchings' California Magazine* focused on California's natural wonders. The inaugural issue included a third retelling of his visit to the Yosemite Valley, accompanied by three of Ayres's illustrations, captivating readers' imaginations.[18]

Ayers returned to the Yosemite Valley the following summer, producing even more drawings before traveling to New York City, where he exhibited his work at the American Art Union. Impressed by Ayers's artwork, *Harper's Magazine* commissioned the artist to return to Yosemite. Tragically, Ayers drowned on his return trip when his ship capsized in a storm off Point Dume. Yet other artists, including Albert Bierstadt, Thomas Hill, and William Hahn, soon followed Ayers to Yosemite, capturing its landscape on paper and canvas. Each artist sparked the public's imagination with their depictions of the valley and its granite peaks, drawing even more curious explorers to the remote mountain valley.[19]

Hutchings led a second expedition into the valley four years later. This time he invited photographer Charles Leander Weed, who brought his forty-pound wood-and-metal camera. Weed took many of the first photographs of the valley, forever altering how visitors saw its grandeur. He took full advantage of the Yosemite Valley's scale and sublimity, producing stunning photographs. As he had with Ayers's drawings, Hutchings published Weed's photographs in *Hutchings' California Magazine*. And, like Ayers's, Weed's images of Yosemite reached far beyond California, with galleries in New York and Paris showcasing his work, bringing both the photographer and the Yosemite Valley international acclaim. Weed soon left California, traveling across the Pacific, and he photographed Hong Kong, Shanghai, and much of Japan. But it would be another photographer that would bring wide acclaim to Yosemite through the lens of his mammoth camera.[20]

GALEN CLARK STOOD motionless at the base of the Grizzly Giant. Heavily bearded, with a well-worn rifle in his hands and a leather satchel slung over his shoulder, he gazed into the distance. A few feet away, Carleton Watkins briskly instructed Clark to hold the pose before ducking under the canvas hood of his camera. It had taken much of the day to set up the shot, with Clark and Watkins walking through the grove of giant sequoias to find the perfect spot. This was not Watkins's first time photographing the sequoia grove; four years earlier, he had captured the very tree Clark now stood beside with his enormous handmade camera. That earlier photograph of the Giant Grizzly, named for its impressive size, had captivated audiences nationwide. But this new photograph would tell a different story.

Watkins emerged from beneath his camera's hood and hurriedly carried the massive glass plate to a small black tent. There, he expertly poured chemicals evenly over the plate before rinsing it with water to halt the chemical process. He then placed the eighteen-by-twenty-two-inch plate into a second chemical bath, permanently fixing the image of Clark onto the glass. Once that was done, Watkins, or one of his assistants, varnished the plate before carefully storing it in a wooden box. He would later print the image in his San Francisco studio, but for now, the day's work was complete, and the group returned to Clark's rustic lodge.[21]

The photograph further propelled each man's rising national fame. Clark had settled along the banks of the South Merced River nearly a decade earlier, opening a modest lodge and guiding tourists to both the Yosemite Valley and Mariposa Grove. A lifelong sufferer of tuberculosis, the forty-four-year-old Clark had endured a difficult life up to that point. Born in rural Quebec, he moved to Missouri in 1836, where he met and married his neighbor Rebecca McCoy. Yet Clark struggled as a farmer and often had to borrow money from his brother to make ends meet. After the tragic death of his wife during the birth of their fifth child, Clark left his remaining children with relatives in New Hampshire and fled to California's gold fields, hoping not only to find his fortune but also to escape to a healthier climate.[22]

Clark was disappointed by the crowds of dejected miners and the gloomy winter weather when he arrived in San Francisco months later. He soon found work with a government survey crew and, after a year of mapping the western foothills of the Sierras, briefly tried gold prospecting before finding a position with the Mariposa Ditch Company. Intrigued by Hutchings's account of the Yosemite Valley, Clark joined a small group of tourists traveling to the valley in the spring of 1856. He was quickly captivated by its beauty and the surrounding mountainous terrain. The party spent several days exploring the region before returning home.

Following his trip to the Sierra, Clark contracted tuberculosis yet again and was given just six months to live. Crestfallen and a continent away from his family, his thoughts turned to a meadow high in the Sierra Nevada that he had glimpsed while traveling to the Yosemite Valley. Determined to take his chances of either dying or recovering in the mountains, Clark moved to his meadow. There, he built a cabin on the banks of the South Merced River and began exploring the territory. Within a year, Clark had recovered and constructed a more substantial cabin, and he began hosting tourists on their way to the Yosemite Valley. It is not certain how many visited Yosemite in those first years, but a decade later, in 1864, James Hutchings counted only 147 visitors to the valley. A handful of those enjoyed Clark's hospitality. Generous to a fault, perhaps because of his brush with death, Clark quickly became known for his meals of wild game and tales of adventure in the High Sierra around the campfire. Virtually every notable tourist to the region, from Frederick Law Olmsted to Ralph Waldo Emerson, stayed at Clark's Station, often sleeping in canvas tents under the ponderosas.[23]

Yet it was another grove of trees that brought Clark fame. He had grown curious about rumors of a grove of giant sequoias hidden in a nearby hollow. A similar grove had already begun attracting sightseers further north, near Calaveras. Although not as tall as the redwoods found throughout the state, sequoias were far more magnificent. Reaching over three hundred feet in height, these 3,500-year-old trees defied imagination. So much so that many viewers believed the bark taken from one of the Calaveras trees and displayed during the Crystal Palace exhibition in London to be a hoax, believing that no tree could possibly be so large. If another grove existed, Clark wanted to see it. One morning, he joined Milton Mann, who would later build the toll road from Clark's Station to Inspiration Point, and set out to find the grove.

Neither man recorded his first experience rambling among the giant trees, but Clark realized that the grove would attract tourists on their way to the Yosemite Valley. He soon began writing stories about the trees' majesty for local papers in hopes of drawing customers to his modest lodge. Visitors did come, and Clark's name quickly became synonymous with the grove. Although neither Clark nor Mann was the first to walk among the sequoias, or even the first white settler to do so, Clark became celebrated as the Mariposa Grove's "discoverer." Much like Bunnell's account of stumbling upon the Yosemite Valley, the tale of Clark's "discovery" erased the presence of the region's Native inhabitants, replacing their rich, complex stories with a fresh narrative that insisted on the land's discovery anew, as if it had never been known, cherished, or lived upon before. Clark leveraged his growing fame by guiding countless visitors through the grove and hosting them at his simple lodge. Appointed guardian of the Yosemite Grant in 1864, he remained in Yosemite until his failing health forced him to move to his daughter's home in Oakland, where he died in 1910 at the age of ninety-seven.

Like thousands of others, Carleton Watkins came to California in search of wealth and, like most, failed at prospecting. Dejected, he turned to a series of odd jobs in and around San Francisco to make ends meet, eventually landing a position at a photography studio in San Jose, where he took pictures of infants. It is unclear how or when Watkins learned photography, a highly detailed endeavor at the time, but by 1856 his career as a photographer was well underway. That summer, investor Henry Laurencel hired Watkins to take photographs for a lawsuit over the ownership of the Fossat Quicksilver mine. The dispute centered around the location of an oak tree that Mexican officials had used as the mine's boundary marker. With limited experience, Watkins traveled to the mine and took several panoramic photographs that were later used in the court's proceedings.

Watkins's photographs caught the eye of Frederick Billings. Already a wealthy man, the minority stakeholder in the Las Mariposas gold mine believed he could become one of the richest men in the country if he could wrest ownership of the mine from its owner, John Fremont. To achieve this, he needed to attract European and Eastern investors. Despite Watkins's limited experience, or perhaps because of it, Billings hired the novice photographer to take several photographs of the property for marketing purposes. The commission not only transformed Watkins's future but also had a significant impact in shaping California's future.[24]

Politics was never far from anyone's mind during the summer of 1860, when Watkins arrived at Los Mariposas. The looming national election and growing concerns about California's future were prominent topics of conversation throughout the state. Despite these worries, Watkins remained focused on his task. Few had undertaken such a significant photography project before, and Watkins had come prepared with hundreds of pounds of equipment and chemicals. Over the following weeks, he and his assistants hiked the hills around Los Mariposas, taking photograph after photograph of the ranch's landscape and mining operations. The experience proved invaluable, allowing Watkins to master his craft and establish his reputation as a photographer.

Watkins took dozens of photographs of Los Mariposas that summer. But instead of focusing on a single element, as was the fashion at the time, Watkins composed larger landscapes through the lens of his camera. His now famous *View from Mount Josephine*, for example, contrasts a road cutting across a hillside with the distant Merced River, creating a ribbon of beige and silver that guides the viewer through the entire terrain. The composition highlights both the natural landscape and the developments already constructed by the mine. This juxtaposition of the man-made and the natural world captured the West's beauty and was a technique Watkins would employ again in Yosemite.[25]

Watkins returned to a rapidly changing San Francisco to finish processing his photographs that summer. The arrival of Boston minister Thomas Starr King had set the city abuzz, with hundreds flocking to the First Unitarian Church on Stockton Street every Sunday to hear the renowned orator preach. Among the attendees was thirty-six-year-old Jessie Fremont. An ardent abolitionist, she had managed her husband John Fremont's presidential campaign four years earlier before the couple moved to their California ranch in the Central Valley. Seeking community, Jessie and her children relocated to a smaller farm named Black Point, overlooking the Golden Gate, that same summer. The farm became the center of San Francisco's political and intellectual life. She often hosted artists like Watkins, writers like Bret Harte, and eventually even Starr King for salons to discuss literature, art, and the day's pressing issues. Politics was a frequent topic of conversation at Black Point, with discussions often focusing on preventing pro-slavery Democrats like William Gwin from dividing California. Yet politics was not the only topic of conversation. Art, literature, and the wonders of the Yosemite Valley were also prominent topics at Black Point, particularly for Starr King.

After journeying from New York, traversing the Panama isthmus, and then sailing up the Mexican and Californian coasts, Thomas Starr King and his wife, Julia, finally reached San Francisco in April. They had been seen off months earlier by their longtime friend and *New-York Tribune* editor Horace Greeley, who was in the midst of publishing a book chronicling his overland voyage to California. In his book, Greeley extolled the state's beauty with particular emphasis on the Yosemite Valley's grandeur. As the two friends had parted ways, Greeley had urged Starr King to make every effort to visit the remarkable valley. Curious, Starr King pledged to see Yosemite before embarking on the next leg of his journey. However, upon his arrival in the Bay Area, Starr King found himself quickly absorbed in church and social obligations, delaying his visit to Yosemite until the following July.[26]

Like most San Franciscans who made the trip to the Yosemite Valley, Starr King ventured inland first by steamship, then by stagecoach, before eventually arriving at Clark's camp, where the two men quickly bonded over their shared love of the New Hampshire mountains. The following morning Clark led Starr King and his fellow sightseers to the Mariposa Grove. Growing tired of measuring each tree's width, Starr King wandered off to nap beneath one of the giants. "I lay for half an hour alone at the root of the most colossal bole, my companions out of sight and hearing, and watched the golden sunshine mount the amber trunk, leaving a hundred feet in shadow and flooding the mighty boughs and locks with tender lustre [*sic*]," he later wrote on the view through the sequoia's branches. Lying there, Starr King pondered how many centuries the tree had stood in such silence. Such thoughts gave the minister pause.[27]

While overwhelmed by the grove's magnificence, Starr King observed that the stalwart trees stood "unprofaned except by fire," just as the Creator had fashioned them. Greeley had made a similar observation the previous summer, noting that fire had scarred, gorged, and hollowed out the ancient trees, which he saw as a sign of the grove's imminent extinction. Such assessments of wildfire's threat were common among non–Native American visitors. Many blamed local Native tribes for setting such blazes, mistakenly viewing the Miwok and Paiute use of fire as destructive rather than crucial to the region's ecological health. Instead, visitors and settlers increasingly sought to suppress all fire in both the grove and the valley, believing that removing fire was beneficial to the natural environment and essential for preserving the landscape that so many came to experience.[28]

Fear of fire became so ingrained in the public's mind that after a blaze swept through the Mariposa Grove in 1889, critics cited the fire as conclusive evidence of the state commission's negligence in protecting the grove. For decades, fire suppression remained the primary policy in the Sierra Nevada, leading to catastrophic consequences. The region's meadows and forests, particularly the giant sequoia groves, relied on fire to clear out smaller trees and grasses that competed with the sequoias for water and nutrients. The removal of fire inadvertently endangered these giant trees. It was not until the 1960s that Park Service scientists recognized the scale of this mistake and started reintroducing fire into the park, often having to educate visitors who still viewed fire as a threat to the park's preservation.[29]

After an afternoon touring the grove, the group returned to Clark's for the evening. The next morning, they mounted their horses before sunrise and headed to the Yosemite Valley. Upon reaching Inspiration Point, they marveled at the view before making their way to the valley floor, where they sat at the base of Bridalveil Fall, getting soaked in its mist before continuing on to the base of El Capitan and then Yosemite Falls. Awed beyond words, a notable achievement for the loquacious Starr King, the group reluctantly left the falls and made their way to the roughhewn two-story building that served as the valley's only true lodge. There, they ate lobster and oysters, a rare treat in such a remote location.

As he had during his travels in New Hampshire's White Mountains the summer before, publishing them in his book *The White Hills: their Legends, Landscapes, & Poetry*, Starr King chronicled his trip to Yosemite in a series of letters to the *Boston Evening Transcript*. These letters, along with Greeley's account, in his book, of his time in the Yosemite Valley, validated the tales of the valley's towering waterfalls and colossal trees to Eastern readers. Yet, even as Greeley and Starr King's rhetoric lauded the Sierra's sublimity to audiences outside of San Francisco, it would be Watkins's photographs that would truly garner Yosemite national attention.[30]

WAR HAD ALREADY broken out in the East when Carleton Watkins left San Francisco for the Yosemite Valley in the summer of 1861. He had spent much of the winter preparing for the months-long expedition to photograph both the valley and the Mariposa Grove, packing several wagons with his camera and needed supplies. After arriving at summer's peak, he spent weeks capturing the valley's granite massifs with his new, massive camera. More than twice the size of the camera he had used in photographing Los Mariposas, Watkins's new camera utilized enormous glass plates he called mammoth plates. Such plates allowed him to capture expansive landscapes and produce prints at a much larger scale. Transporting such large, delicate plates proved exceptionally challenging, however, especially when traveling through the Sierra Nevada. Nevertheless, Watkins photographed at least a dozen different sites within the Yosemite Valley and two in the Mariposa Grove.[31]

It took Watkins nearly a year to print his Yosemite photographs. But by the following fall, he finally crated and shipped a series of large photographs to the Goupil art gallery in New York City. The *New York Daily Times* praised the exhibit, observing, "The views of lofty mountains, of gigantic trees, of falls of water which seem to descend from heights in the heavens and break into mists before they reach the ground, are indescribably unique and beautiful." Standing in front of Watkins's immense photographs, audiences could experience standing along the banks of the Merced River, gaze at El Capitan, or stare at Half Dome's sheer granite wall from the valley floor, without traveling to the remote mountain valley on the other side of the country. So moved upon seeing the images, famed landscape painter Albert Bierstadt traveled to California to paint Yosemite's majestic valley the following year. Starr King sent stereographs of the images to his friend Oliver Wendell Holmes Sr. after the poet requested pictures of the giant sequoias. Starr King also sent several large Watkins prints to Ralph Waldo Emerson, one of which remains on display in the author's Concord home. But perhaps most notably, Senator John Conness presented President Abraham Lincoln a series of Watkins photographs in efforts to convince the President to sign a bill ceding the valley and grove to the state of California.[32]

THOSE INVOLVED IN the campaign to cede the Yosemite Valley and Mariposa Grove left little record of their efforts. A single letter from Central American Steamship Transit Company representative Israel Ward Raymond to California's junior Senator John Conness in February 1864 is the only documented evidence of the idea's origins. In the letter, Raymond urged the senator to introduce legislation to transfer both the valley and the grove to the care of the state, arguing that the federal government needed to act swiftly to prevent them from falling into private hands. Raymond provided a brief description of the region and emphasized that the Yosemite Valley and Mariposa Grove should be "granted for public use, resort, and recreation and be inalienable forever, although leases may be granted for portions not to exceed ten years." He also proposed that all income derived from such leases be used for the preservation of the valley and grove and for the improvement of roads into the region.[33]

Moreover, the letter alluded to previous conversations among "gentlemen of fortune, of taste, and of refinement" who were willing to manage the valley and grove if the federal government ceded guardianship of both to the state government. Historians have long inferred that this group included some of California's most politically influential individuals, many of whom gathered at Jessie Fremont's Black Point salons. This group likely included Thomas Starr King, Carleton Watkins, Israel Ward Raymond, and Frederick Law Olmsted, to name but a few. Tragically, Starr King died from a combination of diphtheria and pneumonia brought on by exhaustion a month after Raymond presented his letter to Senator Conness. Yet it was Olmsted, whose influence is evident in the letter's key passages, who would ultimately refine the Yosemite Grant's purpose.

Olmsted was already a national figure when he moved his family to California in 1863 to manage the John C. Fremont's Los Mariposas Estate. His writings on British gardens and the American South earned him a national reputation as a prominent social reformer. More important to his new task, Olmsted was also an extraordinarily effective administrator, an ability he had honed while collaborating with English-born architect Calvert Vaux in building New York's Central Park.

The term "park" entered the American popular lexicon in the decades leading up to the Civil War. Inspired by the well-manicured landscapes of European city parks they had enjoyed while traveling, wealthy Americans sought to create their own grand urban parks to demonstrate America's emergence as a cosmopolitan nation equal to Europe. New York's Central Park emerged from this desire, reflecting the aspirations of affluent New Yorkers to elevate their city to the status of London or Paris. Additionally, there was a push to replace New York's relatively small ten-acre Battery Park, which many deemed overcrowded with "loafers" and insufficient for the city's growing population. Various solutions were proposed, including expanding Battery Park into the harbor, but city officials ultimately decided to construct a large central park on New York City's northern edge.[34]

In 1853, the New York state legislature authorized the city to claim seven hundred acres in the center of Manhattan Island. That as many as 1,600 called the area home was of seemingly little concern to the park's wealthy advocates. Five years later, the newly appointed Central Park Commission announced an open competition for the park's design. Thirty-three submitted plans for the proposed park. All included a parade ground, formal garden, major fountain, and concert hall and integrated the city's reservoir, but they differed on how to best integrate each element into the landscape. After deliberating over numerous submissions, the Commission finally settled on landscape architect Calvert Vaux and the park's superintendent Frederick Law Olmsted's Greenwald Plan.

Vaux and Olmsted's partnership was more political than professional. The British-born landscape architect needed Olmsted's connections as the park's superintendent to secure the contract to design Central Park. Although born into wealth, Olmsted had worked in various roles, including as an apprentice seaman, a merchant, a farmer, a journalist, and Central Park's first superintendent, by the time he met the British architect. The pair's Greenwald Plan evoked an idyllic pastoral retreat, inspiring the public with its natural beauty. They envisioned a space free of roads, instead creating sweeping pathways for carriages, horses, and pedestrians around its perimeter, shielding the park from the city's hustle. To maintain this serene environment, Olmsted employed two dozen park keepers to prevent visitors from carving their names into trees, racing horses and carriages around the pathways, or walking on the grass. Educating a diverse range of visitors on how to properly appreciate the park was a continual challenge, as many resisted the rules intended to preserve the landscape for contemplation rather than entertainment. Nevertheless, Olmsted remained committed to the idea of nature as a source of respite and reflection, later making it central to his vision for the Yosemite Valley.[35]

Seeking to contribute to the war effort, and increasingly at odds with Central Park's commissioners, Olmsted resigned as the park's superintendent to become the general secretary for the United States Sanitary Commission at the start of the Civil War. The Commission's primary task was to care for the wounded, a harrowing responsibility in any war, but especially in a conflict where the scale of destruction far surpassed the strategies of generals. Exhausted by the carnage and a prolonged political fight over managing the Commission, Olmsted resigned his post after two years. Unable to return to his work in building Central Park, and in need of income, he accepted a job managing John and Jessie Fremont's Los Mariposa Estate.[36]

Olmsted's move to California came at a pivotal moment in the nation's history. The Civil War continued to drag on, becoming seemingly bloodier with each passing month. With a presidential election approaching, the Lincoln administration was increasingly reliant on both California's electoral votes and resources. At the same time, concerns about private development within the Yosemite Valley spurred efforts to protect the valley from private interests by lobbying the federal government to cede the valley and adjacent Mariposa Grove to the state of California.

Precedent for the federal government ceding land to states already existed. The Morrill Act of 1862 granted land to states in order to fund land-grant universities despite Southern Democrats' opposition that it would benefit free-soil settlers and unconstitutionally expand federal power. Democrats had also opposed the Homestead Act of 1862 for similar reasons. However, the South's secession had created an opportunity for the Republican Party to cede federal land. Ceding control of the Yosemite Valley and Mariposa

Grove to California as a favor to those advocating for its protection, as well as helping secure Lincoln's reelection, was a politically prudent move, proponents argued. That it cost the federal government only a few acres of "worthless" land made it all the more attractive to a Congress concerned with ending the war.[37]

Conness quickly acted on Raymond's letter. After requesting further information from the General Land Office, the senator submitted a bill closing further land sales in—and entrusting the guardianship of both the Yosemite Valley and the Mariposa Grove to—California. The bill incorporated much of the letter's language, chiefly stipulating that the lands be "held for public use, resort, and recreation." It also provided for recreational amenities to be managed by private interests through leases not exceeding ten years. This arrangement was designed to ensure visitors could enjoy the area while the state protected it from commercial overdevelopment. Both the valley and the grove would remain federal land, but California would manage both. The bill received little debate in either the House or Senate, and on June 30, 1864, weeks after General Ulysses Grant's loss to General Robert E. Lee at Cold Harbor, President Lincoln signed the Yosemite Act into law.

The Yosemite Act marked a significant milestone in the development of the U.S. National Parks. The concept of a national park had gradually evolved over the previous three decades. In 1832 President Andrew Jackson designated Hot Springs in Arkansas as the first national park, but Congress never passed legislation to manage the site, and it eventually fell into private hands. Despite this setback, the idea of national parks persisted. That same year, artist George Catlin proposed creating a vast park across the Great Plains to protect its immense bison herds and the Indigenous peoples living in the region. But unlike Hot Springs, Catlin's vision aimed at preserving Indigenous ways of life, not the enjoyment of nature. Catlin's vision never came to fruition as America's hunger for land overwhelmed any calls to preserve the Great Plains and its peoples. Nonetheless, the seeds of the national parks had been planted by the Civil War's end.

TEMPERATURES HAD RISEN well into the nineties by the end of June 1864, as Harriet Errington sat down to write a letter to her sister Lottie, who was now living in England with her husband and growing family. It was a far cry from the two sisters' childhood on Staten Island. No rain had fallen for days, and while the evenings were pleasant, Harriet and the children were forced to retreat inside most afternoons to escape the midday heat. The Los

Mariposas ranch was a long way from the Atlantic Seaboard and her friends, Errington lamented. "Still, I am glad I came here so far, for the agreeable has been greater than the disagreeable," she wrote. The governess and tutor to the Olmsted's six children, she had followed the family from New York to the California goldfields. The ranch was a pleasant place, she reported, with celebrities like Starr King always passing through on their way to or from touring the Yosemite Valley and grove of giant trees.[38]

The Olmsteds had yet to see the valley and grove. Keeping the mine solvent had dominated much of Frederick's days, with evenings spent with the family touring the hillsides above the ranch, picking flowers, and watching California's magical sunsets. By the fourth of July 1864, he decided it was time for the family to escape the increasingly oppressive heat and visit the High Sierra. And so ten days later Olmsted, his wife, Mary; their six children; their Black guide and servant, known simply as Bell; and Errington headed out from their ranch to Clark's Station. Arriving two days later, the family set up camp alongside the South Fork of the Merced River.[39]

Errington had a small tent to herself within the camp. She slept well, wrapped in heavy wool blankets, though she still found the mornings a bit chilly. "This is my first experience of tent life, and I find it perfectly congenial," she later reported to her sister in a letter on the trip. One evening, after a day of lessons and sunbathing by the river, the family ate bear for dinner. A local hunter had killed the beast and given it to Clark to serve his guests. After tasting the greasy meat, Harriet concluded that bears ought to be shot but not eaten. She continued to guide the children through their French lessons over the coming days until Mary Olmsted declared there was no need for further school until they returned home. Freed from teaching, Harriet joined Frederick in exploring the Mariposa Grove. After riding for five miles up a steep trail carpeted with pine needles, the two suddenly came upon the trunk of a majestic sequoia. "The great beauty of these forest kings is as striking as their size," she wrote. Entering the grove, they wandered among the giant trees, measuring one tree's trunk at fifty feet in diameter.[40]

So went the next three weeks. The family spent their days exploring the Mariposa Grove, idling in camp, or riding farther afield. A small fire near the grove caused some excitement, but not enough to move the camp. Having left for a week for some business at the mine, Frederick returned to camp and announced it was time to venture to the Yosemite Valley. The next morning the group reloaded the mules, wagons, and horses, thanked Clark for his hospitality, and headed down the narrow trail to the valley.

Frederick and Mary rode ahead, leaving Errington and Bell to bring the children along at a more casual pace. After spending the night at Harvey Ostrander's cabin at the junction of the Glacier Point and Old Mono trails, Errington, Bell, and the children arrived somewhat late at Inspiration Point. "The sun was near setting and the rock walls of the Valley were bathed in its light above and lost in the gloom below," Errington wrote of the view that evening by the fire in another letter to her sister. In it, she also implored Lottie to travel to see the Watkins photographs now on exhibit in London as soon as she could in order to understand the valley's beauty. "He is here now taking new beautiful views and seems so lukewarm about pushing their sale that I am afraid it will be long before they become known," Errington told her sister.[41]

Arriving later at the Olmsteds' camp at the base of the Yosemite Falls with the children, Errington quickly set out to explore the valley. It was far more wonderous than she had anticipated. Half Dome enthralled her, as did the Yosemite Falls as it spilled from one ledge to another before cascading into a pool at its base. Enjoyment of the scenery was the family's principal occupation over the next several days. Olmsted certainly spent some time meeting and talking to the few who had settled in the valley, including James Hutchings. After another hasty trip to Bear Valley, Olmsted joined California Geological Surveyor William Brewster in exploring the High Sierras, before the family finally returned home in early September.[42]

Shortly after the Olmsteds' Sierra vacation, California Governor Frederick Low appointed Olmsted as the nominal chairman of the newly formed Yosemite Commission. Composed of five prominent state and local businessmen and politicians, including the chief of the California Geological Survey, Josiah D. Whitney, and Israel Raymond, the Commission was tasked with overseeing the management of the newly created state park. Olmsted was a natural choice to lead the group. Having witnessed the horrors of the Civil War, the urban decay of New York, and the injustices of slavery, he strongly believed that the nation could find healing in nature's sublimity. And that, like Central Park, Yosemite offered the promise of that respite.[43]

Olmsted spent the next year reflecting on the Grant's purpose and drafting a report for the Commission on how best to manage California's new charge. The family's vacation the previous summer had confirmed much of what he had read about the grove and valley's beauty, as well as his concerns about private interests potentially seizing control of both. He reasoned that the Yosemite Grant should be made accessible for all to enjoy, especially in light of the horrors he had witnessed during the war. The Grant then, he believed, could serve as a place for contemplation and renewal. With these ideas in mind, he began drafting a report outlining his vision for Yosemite's future.

On August 9, 1865, Olmsted stood in the shade of the Yosemite Valley's ponderosa pines and presented the report to his fellow commissioners. He emphasized the restorative qualities of quiet contemplation in the presence of natural wonders like Yosemite Falls and Half Dome, arguing that the Yosemite Grant must be reserved for "the occasional contemplation of natural scenes of an impressive character," which "not only gives pleasure in the moment but enhances the subsequent capacity for happiness and the means of securing it." He asserted that standing at the base of a giant sequoia or gazing into the Yosemite Valley from Inspiration Point offered relief from daily worries and would increase one's capacity for future happiness. Yosemite's contemplative nature, he argued, should remain its primary purpose, and any commercialization or alternative uses be deemed inappropriate. Olmsted contended it was the government's duty to protect such places from those who sought to profit at the expense of the public's enjoyment. Such actions were undemocratic, he argued. The Commission's charge then was to ensure that "the mass of the people" could benefit from the unique features of the Grant, which had led Congress to treat it differently from the rest of the public domain.[44]

Olmsted proposed two key considerations for preserving the Grant. First, he advocated for valuing the Grant in its present condition as a living museum of natural science. He drew a parallel to the disappearance of native plants along the Atlantic coast, warning that Yosemite could face a similar fate if the Commission did not act soon. Olmsted observed that "many of the finer specimens of the most important trees in the Yosemite scenery have already been destroyed." He noted the damage caused by felled and girdled trees, fires set by both visitors and Indigenous people, and graffiti on picturesque rocks throughout the valley. Without intervention such destruction would continue, he warned.

Second, Olmsted highlighted the grim reality that tourism often leads to overdevelopment and overcrowding. Citing the Alps and New Hampshire's White Mountains as examples, he cautioned that allowing development within the valley could result in overcrowding, warning, "The slight harm which the few hundred visitors of this year might do, if no care were taken to prevent it, would not be slight if it were repeated by millions." While Yosemite should be accessible for all to enjoy, he argued, its beauty must also be preserved to ensure that it could be enjoyed equally by future generations.[45]

With the war over, Olmsted felt the pull of home. So he returned to New York City, where he rejoined Vaux in completing Central Park before going on to design hundreds of city parks across the country, becoming the nation's most renowned landscape architect. His report on Yosemite was forgotten until its rediscovery in 1952 by his biographer, Laura Wood Roper. Speculation suggests that his fellow commissioners may have conspired to suppress the report, fearing that its recommendations might divert state funds from the California Geological Survey. Whatever the reason for the report's disappearance, the Commission did not heed Olmsted's warnings. Lacking state support, they became increasingly reliant on private interests to provide lodging and amenities within the Grant. This led to significant degradation within the valley. Trees were felled for lumber, the moraine was dynamited to its drain its wetlands, and the suppression of Miwok fire practices allowed undergrowth to build. Within a decade, criticisms of the Commission's negligence led to their removal and sparked calls for the establishment of a national park.

NEITHER JAMES HUTCHINGS nor James Lamon had much to celebrate following Congress's passage of the Yosemite Grant Act. By ceding the Yosemite Valley and Mariposa Grove to California, Congress had effectively nullified each man's claim to his property within the valley. Despite the long-standing practice of preemption, which recognized squatters' rights once a government survey established boundaries and entered the land into the public domain, neither Lamon nor Hutchings retained title to his property under the act. Lamon had moved to the Yosemite Valley in 1859 and developed 219 acres by planting several hundred fruit trees, establishing several large gardens, and maintaining two pastures. Similarly, Hutchings had purchased the two-story Upper Hotel and relocated his family to the valley just months before its cession. Despite their investments and improvements, both men and their families were now considered illegal squatters, with no valid claim to their properties.[46]

In truth, private land claims like those of Hutchings and Lamon presented a significant challenge to California's ability to manage the Grant. Private interests threatened to not only undermine the state's authority within the Grant but also compete for tourist dollars the state needed to develop roads and trails in Yosemite. To address this issue, the Yosemite Commission offered both men, as well as the owners of the Leidig Hotel, ten-year leases to operate businesses on their claims. Outraged by what he saw as a miserly offer,

Hutchings rejected the Commission's proposal outright. Lamon also declined the state's offer, citing his long-standing investment and tenure in the valley.[47]

The conflict over preemption essentially reflected competing visions of Yosemite's future. Josiah Whitney, chief of the California Geological Survey, expressed concern that if the Yosemite Valley fell into private hands, rather than being managed by the state for the public's benefit, it would be exploited for private gain rather than serving the common good. He warned that the valley could become, like Niagara Falls, a "gigantic institution for fleecing the public," rather than a place of enduring beauty and joy. In this narrative, Hutchings was cast as the villain, portrayed as greedy and solely interested in his own self-promotion. While this portrayal had some basis in Hutchings's attempts to profit from the valley through his magazine and hotel, it also reflected broader tensions of the era over land ownership and use throughout the American West.[48]

For decades, settlers' claims of ownership were rarely questioned, even if they had illegally squatted on an unsurveyed federal domain. And few questioned the theft of Native treaty lands. Hutchings faced the challenge of powerful political forces that sought to protect the Grant for public use, recreation, and preservation. These groups aimed to prevent the rampant commercialization that had overtaken Niagara Falls. For many reformers, this commercialization not only marred the site's beauty but also undermined its significance. Critics argued that the falls had become choked with vendors selling cheap souvenirs and hucksters charging unwitting tourists for views of the falls from its edge. Reformers hoped to keep such crass commercial exploitation from infecting the Yosemite Valley. Such criticisms often ignored tourism's devil's bargain, that tourists' enjoyment of a place's natural beauty often changes the very nature of that place by leading to its development. But the debate was more than whether preservation or commercialization should be held supreme within Yosemite, but how visitors experienced its beauty through their own recreational enjoyment.[49]

Lamon eventually agreed to sell his house and orchard to the state for $12,000. The owners of the Lower Hotel and Leidig Hotel likewise sold their claims but continued to run their small inns. Hutchings, however, rejected the state's offer and continued his fight in court. The case eventually made it to the U.S. Supreme Court, which, in 1873, affirmed the federal government's right to withhold land from the public domain for the public's enjoyment. This decision extinguished Hutchings's claim and set a precedent for the establishment of future national parks.[50]

Embittered by the court's ruling, Hutchings accepted the state's payment and moved to San Francisco. Ironically, in 1880, after the state legislature dismissed members of the Yosemite Commission over charges of corruption, it hired Hutchings to replace Galen Clark as the Grant's guardian. The irascible booster held the position for the next four years, until, amid ceaseless complaints about his insolence, the state reversed its decision and reinstated Clark. In 1902 Hutchings died while vacationing in Yosemite when his wagon crashed on the Big Oak Flat Road. His legacy remains elusive through today. While Hutchings was among the first to reveal the Yosemite Valley to the world, his unrelenting boosterism and self-promotion foreshadowed the coming debate over Yosemite's development.[51]

HAVING JUST RETURNED from spending two weeks exploring the High Sierras, John Muir sat down to write his sister Sarah. It had been a difficult but rewarding journey, he explained. He had slept among the mountain peaks with no blanket and little to eat. Even though he had just returned, he would be heading out again soon. South this time to continue his research along the Kings River, then on to Lake Tahoe before returning to Oakland to write through the winter. "The mountains are calling, and I must go, and will work on while I can, studying incessantly," Muir wrote, ending the letter by sending his best wishes to the rest of the family.[52]

A year had passed since Muir had published his first article, in the *New-York Tribune*, on how glaciers had shaped the Sierra Nevada. The short piece marked the start of a writing career that would span the next four decades and transform American environmental thought and politics. Other articles soon followed. The next was a piece on Yosemite in winter, then, months later, a companion story on Yosemite in spring. In 1873 Muir wrote his first article on the Hetch Hetchy Valley, this time for the *Boston Weekly Transcript*. By the end of the decade, he was contributing to *Harper's New Monthly Magazine* and the *San Francisco Daily Evening Bulletin*. Lyrical in style, each of Muir's pieces drew readers into his Sierra—a world of awe and wonder where even the smallest detail would engross them for pages. This distinctive style would become one of Muir's hallmarks as his body of work and his fame continued to grow. However, when he wrote to his sister from the Yosemite Valley that July, he was years away from becoming the renowned sage of the Sierra, his name synonymous with Yosemite.[53]

Muir had arrived in San Francisco in the spring of 1868. After spending a day in the city, he approached a carpenter and asked for directions out of town. When the man inquired where he wished to go, Muir responded, "Anywhere that is wild." Soon, he was on the Oakland Ferry and heading toward the Yosemite Valley. Details of his first visit to Yosemite, however, are scant. Muir's journal entries ceased as he crossed the Panama isthmus and did not resume until a year later. He briefly mentioned his inaugural trip to Yosemite in a letter to his mentor and close friend Jeanne Carr, in July of 1868. However, the exact timing of his arrival and the duration of his stay remain lost to history.[54]

Muir returned to the Sierra Nevada that following summer, finding work as a shepherd and meticulously recording his observations of every rock, tree, bear, and insect as he and his fellow shepherd guided their flock through the mountains, circling the Yosemite Valley to Lake Tenaya before returning to Brown's Flat. At the season's end, he returned to the Yosemite Valley to work for James Hutchings. Over the next year, the thirty-one-year-old Muir worked for Hutchings's building, and then operated a sawmill, and he guided visitors through the valley, most notably the writer Ralph Waldo Emerson.

News of the renowned author's planned visit to Yosemite in the spring of 1871 elated Muir. He quickly sent a letter offering to show Emerson Yosemite's wonders. Intrigued by the letter's enthusiasm, Emerson agreed to meet with the eccentric naturalist, and the two met at Muir's sawmill. The ecstatic Muir excitedly toured Emerson through the site, showing him the improvements he had made in constructing the mill. Amused by the tour, Emerson invited Muir to join his entourage to the Mariposa Grove the following day. Muir readily agreed and, the following morning, rode alongside the older man as the group traveled to the grove, pointing out the natural wonders they encountered on their way to Clark's Lodge. After spending a night at the lodge, the group wandered among the sequoias, measuring their girth and gazing up into their towering canopies. Having guided the group, Clark invited Emerson to name the giant tree next to his cabin. Emerson agreed, naming it Samoset after the first Abenaki sachem who welcomed the Pilgrims at the Plymouth Colony.

Enthralled by Emerson's presence, Muir desperately bade the author to spend his last night in the Grant sleeping under the canopy of the giant sequoias. Emerson's companions spurned Muir's invitation, citing the elderly author's age and health. Instead, they mounted their horses for the ride back to Clark's. Muir walked to the grove's edge to watch Emerson, whom he would later equate to the ageless sequoia, ride off into the distance. Riding at the rear, Emerson paused and turned, and taking his hat off he waved it in goodbye to Muir before finally turning away. It was the last time the two men saw each other, a fact that Muir lamented for the rest of his life. But the two remained in contact, corresponding with each other until Emerson's death. Emerson proposed Muir move to Concord, writing that he could easily take up the author's mantel. Muir declined, writing the Sierra had become his true and only home. Yet it is hard not to see the torch being passed between Emerson and Muir that afternoon within the Mariposa Grove.[55]

Much has been made of Muir being one of the most impactful transcendentalist writers of his era, even though Muir himself never fully identified as such. Instead, his encounter with Emerson led Muir to read Emerson and other transcendentalist writers more earnestly, integrating their ideas into his own thinking. Muir's writings reflected not only Emersonian ideals of nature and beauty but also his own passion for the wild, unspoiled landscapes of the American West. This blend of place and philosophy mirrored Americans' broader embrace of the West and its romantic landscapes, a shift Muir both influenced and capitalized on as he developed as a writer.

Muir spent the next two years in the Sierra Nevada, writing some, but mostly wandering through the mountains. Finally, in 1873, he moved to Oakland, where he struggled to write his first two books. By the time Muir was forty, his friends Ezra and Jeanne Carr, whom he had studied under at the University of Wisconsin, convinced him it was time to settle down and marry. They had introduced him to, Louisa, the daughter of prominent California physician John Strentzel, years before. Louisa shared Muir's love of literature, but not his wanderlust. Nevertheless, the two were married in 1880 and soon had two daughters. John also partnered with his father-in-law in managing the family's large orchard. The marriage not only provided the stability long lacking in Muir's life but made him a wealthy man, affording him the time to dedicate to his writing.

Even with such stability, Muir did not publish his first book until 1894. Like many of his previous articles, *The Mountains of California* described his adventures in the Sierra Nevada and other mountain ranges, including his first ascent of Mount Ritter. Other books followed, each finding a ready audience fascinated with the West's mountains and Muir's travels. While Muir's growing reputation as a writer garnered him some celebrity, it would be his growing criticism of the Yosemite Commission that would cement his legacy.

NEARLY 3,000 PEOPLE visited the Yosemite Valley the summer following the Supreme Court's 1873 ruling, a notable increase from just a decade earlier, when a mere 147 visitors toured the area, according to the Yosemite Commission. The surge in visitation can be attributed to improved access to the valley by both train and stagecoach, as well as California's growing population. The completion of the Transcontinental Railroad also significantly reduced the travel time between New York and San Francisco, allowing tourists from around the world to see Yosemite. Yosemite had become a destination for European and American tourists alike. Travelers no longer had to first sail to Panama, cross the isthmus, and then sail to San Francisco. Instead, they could simply board a train and travel across the Great Plains, through the Rocky Mountains and the Great Basin and over the Sierra Nevada, before traveling by train once again to the Sierra Nevada foothills and then by coach into the valley. Four years later, the Coulterville Road and the Big Oak Flat Road both entered the valley, opening it to stagecoaches for the first time.[56]

Despite such growing interest, the California state legislature resisted providing adequate funding to manage the Yosemite Grant, forcing the Commission to turn to private concessionaires to provide needed lodging and other facilities to meet the growing numbers of tourists. Critics condemned the Commission for such allowances, believing that they would further the Grant's degradation. Public sentiment soon turned against the Commission, and in 1881 the legislature replaced the entire board, including Clark, and hired State Engineer William Hammond Hall to assess the state of the Grant and provide some guidance on its management.

Hall opened his report a year later by asserting the state had a duty to protect the Yosemite Valley from destruction. To accomplish this, he recommended the Grant be expanded to include the larger Merced watershed to keep it from falling into private hands for sheep and cattle grazing, which would in turn foul the Merced River as it flowed through the Yosemite Valley. Second, he argued for the restriction of visitor travel on the valley floor. Unbridled access combined with growing numbers had severely altered the valley's appearance, Hall noted. Continued use at the current levels, he predicted, would surely deteriorate the valley to a point that it would hurt not only the natural resources but also the tourist business. And third, he argued for the necessity to channelize the Merced to preserve meadows needed for livestock feed. "The occupation and use of the water-shed [*sic*] above the valley, the visiting and enjoyment of the valley itself, and the operations designed to prepare for these are all calculated to destroy its attractions," Hall concluded. It was, therefore, essential to acknowledge that meeting the desires of every landowner and visitor would ultimately lead to the valley's ruin.[57]

Yosemite also became a destination for global sightseers. The Commission estimated visitors from twenty-eight foreign countries visited Yosemite in 1885 alone. The majority hailed from Great Britain, followed by Australia and Canada. Among them was John Herbert Roberts. Like many nineteenth-century British travelers, Roberts published his journal of his travels. His *A World Tour: A Year's Diary, Written 1884–1885* took readers from London to New York, onward to Canada, Salt Lake City, Yosemite, and San Francisco, before sailing to Japan, India, and Australia, then home to Liverpool. His inclusion of Yosemite in his grand tour reflected the valley and grove's growing popularity as a destination for foreign tourists.[58]

Roberts arrived in Wawona in the middle of August 1885. Formerly known as Clark's Station, the now bustling community had been purchased by John and Wesley Washburn in 1878. The morning after his arrival, Roberts walked among the giant sequoia before watching men float logs down a flume along a nearby stream. The next day he boarded a coach for the Yosemite Valley as the din from the stables, black smithy, laundry, and lodge filled the air. Over the coming days, Roberts followed what by that point had become a typical itinerary within the Yosemite Valley: first stopping at Inspiration Point to take in the view of the valley before spending the night in a rustic lodge. He watched the sun rise over Mirror Lake, hiked to the base of Vernal and Nevada Falls, and rode a lazy mule to the top of Glacier Point before returning to Wawona and then San Francisco. Like many tourists, Roberts struggled in describing the Grant's beauty, writing instead, "I am afraid I would only weary you, if I tried to describe either the grandeur or the character of the scenery of the Valley." The closest he could come was comparing Yosemite's cliffs to Wales's Snowdon Cut.[59]

Even as more and more visitors like Roberts discovered the Yosemite Grant's beauty, criticisms of the Yosemite Commission's management grew. Articles in the *San Francisco Daily Examiner* alleged the Commission had ignored the destruction of large scenic trees, some more than five feet in diameter, to improve views, and the troublesome amounts of dust raised on the valley floor's stage roads. Others accused the Commission of favoritism, particularly following the Commission's construction of the Stoneman House. The opulent four-story lodge, named after the governor, looked out of place in the valley. The Commission's declaration of Black's and Ledwig's hotels as "unsightly," and its order that they be torn down to draw business to the state-owned Stoneman House, infuriated local business owners, who wrote letters petitioning Congress to investigate the state's management of the Grant.[60]

A fire burned through the Mariposa Grove in the summer of 1889, providing the Commission's opponents with further evidence of the Grant's mismanagement. Settlers had long viewed fire as a threat, nowhere more so than within the grove of giant sequoias. Critics argued that such blazes would devastate the grove, and so the allowance of any fire within the grove ran counter to the Commission's mandate.[61]

However, it was a series of articles by John Muir that drew national attention to the Commission's management of the Yosemite Grant. Through his scientific writings on Yosemite, Muir had befriended *Century Magazine* editor Robert Underwood Johnson. Alarmed by the mounting stories of destruction within the Grant, the two traveled to Yosemite Valley in 1889 to witness firsthand the damage caused by increasing tourism, as well as the effects of sheepherding and timber cutting in the high country. What they found horrified them. Meadows once lush with grass were now overgrazed. Tree stumps sat where groves had once stood. Dust from wagons fouled the air. Most disturbing was the construction of a dam at the lower end of Mirror Lake. Muir followed up on the visit with two articles in *Century* proposing the creation of a national park.[62]

Muir's articles appeared just as Congress began debating the creation of new national park. California Congressman William Vandever had introduced a bill seeking the protection of the Sierra forests. Support by the Southern Pacific Railroad, which envisioned a profitable tourist destination like that in Yellowstone, swayed Congress on the bill. On October 1, 1890, President Benjamin Harrison signed a bill creating Yosemite National Park. The country's third national park, joining Yellowstone and Sequoia, Yosemite remained incomplete. Both the Yosemite Valley and Mariposa Grove remained under state control. While the debate over both remained far from settled, Yosemite was no longer a remote valley far in the High Sierra, but a celebrated destination. Such notoriety would bring its own challenges, including defining what constitutes a national park. However, the foundation of that meaning had been laid by its first generation of visitors. Yosemite had become a symbol of American nature.

2 Making a National Park

Thomas Newsham was appalled at the state of the Yosemite Valley. Fully a third of it was fenced off and under cultivation. Tree stumps lay scattered across the valley floor, stark remnants of the trees felled in order to provide visitors with unobstructed views of Bridalveil and Yosemite Falls. Moreover, the ragtag collection of hotels scattered throughout the valley were little more than rude huts and hovels, operated by people whom Newsham despairingly described as "eyesores" in his first report on the Yosemite Grant to Secretary of the Interior John Noble.[1]

As one of the U.S. Geological Survey's two land inspectors, Newsham had been sent to Yosemite to investigate the mounting allegations of corruption against the Yosemite Park Commission, survey the newly created national park that enclosed both the Yosemite Valley and Mariposa Grove, and provide recommendations on how the Department of the Interior should move forward in managing its newest national park. Alarmed over the rumors of nepotism and fraud, Secretary Noble had grown uneasy at the prospect of becoming entangled in scandal with the California state legislature over the Yosemite Commission's alleged mismanagement of the Yosemite Grant. Needing a sober assessment of the grant's conditions, Noble ordered Newsham to California.[2]

The assignment was a welcomed change from his previous task: reporting on the famine ravaging northern Oklahoma. But despite the stunning scenery in which he found himself, Newsham's trip was anything but a holiday. Tasked with reporting back on the state of both the Yosemite Grant and Yosemite National Park, he spent much of his time recording his impressions of the region. He was horrified at the rampant commercial development that had overwhelmed the grant, but even more distressing was the unchecked logging, grazing, and poaching plaguing the national park. Poachers often killed as many as thirty deer at a time solely for their hides, Newsham reported, while timber interests blatantly felled trees throughout the reserve. He saved his greatest condemnation for the herds of sheep who spent the summers grazing in the Sierra. Writing to the secretary, he noted, "You cannot imagine the utter desolation and untold injury done by the sheep to the grass, flowers, and undergrowth of the mountain valleys of this country." Clearly, something would have to be done to stop such wanton destruction, or soon there would be no park left to enjoy.[3]

After spending his first week exploring the Yosemite Valley, Newsham climbed the Glacier Point Trail to the top of the Sentinel Dome. From its granite-roofed summit, the retired major took in the splendid view of the surrounding peaks and valley below. Citing the lateness of the season, Newsham apologized to the secretary for not making the trek to see the Hetch Hetchy Valley but reported that many argued it was the Yosemite Valley's equal, its cascading waterfalls among the finest in America. Sated by the view, he turned and traveled south to the Chilnualna Falls before making his way to the Mariposa Grove.

Like the hundreds of visitors before him, Newsham fell into a quiet sense of wonder as he wandered through the grove. Nothing could be more beautiful, he reported, than standing at the base of one of the colossal trees. Later, reflecting on his tour, Newsham suggested the inclusion of both the Big Tree Grant and Signal Peak in the newly established national park, foreshadowing the coming battles over the park's boundaries. But there was a much more pressing need than expanding the park's boundaries. "The first and most urgent need for the preservation of the National Yo Semite [*sic*] Park is a strong cavalry force, I would say two companies, to be situated at or near Wawona, a splendid place to camp and the natural entrance to the park," Newsham asserted in his final report to Nobel.[4]

There was a precedent for sending in the cavalry to save the national parks. Just four years earlier, the U.S. Army had assumed command of Yellowstone National Park for very similar reasons. Cavalry troops provided the needed federal presence in patrolling the park, protecting its wildlife and scenic wonders from poachers and vandals. With only fifteen thousand employees, more than a third of which worked for the Pension Service, the Department of the Interior lacked the personnel and expertise to manage even one national park, let alone three. In contrast, the army had both the manpower and authority to enforce federal control within the national parks.[5]

Like Yellowstone National Park before it, Yosemite emerged as a bellwether of Americans' shifting perspectives of Western federal lands. The campaigns that culminated in the creation of Yosemite and Sequoia National Parks were responses to decades marked by speculation and the relentless exploitation of the nation's natural resources. The Yosemite Act, and the 1891 Forest Reserve Act, which followed it, established a precedent of conserving federal lands for public use rather than disposing of them to fill federal coffers. Even with the setting aside of lands as national parks, "national parks" remained ill-defined until nearly the second decade of the twentieth century. That process of making a national park pressed multiple competing interests against each other, concluding with little resolution.

This shift in federal policy was largely driven by California's urban elite, whose burgeoning Progressive political movement sought to manage the nation's natural resources more efficiently through stronger federal oversight. Both Yosemite and Sequoia joined Yellowstone in emphasizing each landscape's recreational value, rather than its timber, mineral, or ranching worth. Despite having now set aside three national parks for the public's enjoyment, the federal government provided no dedicated bureau to oversee the management of these Western lands.[6]

In truth, there was little consensus on what made a park a "national" park. To that end, the Yosemite Act provided minimal guidance on the new park's purpose, only requiring the Department of the Interior to establish and publish regulations for "the preservation from injury of all timber, mineral deposits, natural curiosities, or wonders within said reservation, and their retention in their natural condition." The interior secretary was also permitted to grant leases for small parcels of land to build necessary visitor accommodations.[7]

Moreover, unlike Yellowstone's enabling act, Yosemite's did not define the new national park as a "pleasuring-ground for the benefit and enjoyment of the people." Instead, much of the act's language focused on preventing timber harvesting, grazing, and hunting within the park. But as with Yellowstone, Congress expected Yosemite to be financially self-sustaining and so allocated no budget for the new park. This arrangement quickly proved inadequate in meeting the costs of managing a national park.[8]

Excluding the Yosemite Valley and Mariposa Grove, Yosemite National Park offered few attractions for visitors. Clark's Station and the Wawona Hotel were the only lodging options outside the valley, and a handful of trails connecting Wawona, Glacier Point, the valley, and the High Sierra were the most traveled routes in the region. Many tourists simply passed through the park en route to the giant sequoias or the Yosemite Valley. The majority of the park remained vulnerable to sheep and cattle grazing, logging, and even mining, despite the absence of gold or silver deposits within its boundaries. As Newsham noted in his reports to Secretary Noble, the ongoing trespassing by grazers, loggers, and hunters posed the greatest threat to the park. Yet, without annual appropriations or the necessary administrative structure to enforce the Yosemite Act's mandate, the Department of the Interior once again turned to the U.S. Cavalry.

The cavalry played the foundational role in creating Yosemite National Park. Over the next twenty-six years, cavalry troopers built the park's trails, patrolled its boundaries, enforced its regulations, and defended the park from the encroachment of outside forces. But before they could do all of that, they needed to survey the park. And so, on December 4, 1890, Secretary Noble, acting on Newsham's advice, requested the U.S. Cavalry be stationed in both Yosemite and Sequoia National Parks. That following summer, the Fourth Cavalry rode into the Sierra Nevada to assume command of the country's newest national parks.[9]

THERE IS NO written record of Captain Abram Epperson "Jug" Wood's first impressions of Yosemite. A seasoned veteran of campaigns against the Cheyenne, Ute, and Apache, as well as the Civil War, Wood was likely preoccupied with matters far removed from the landscape's grandeur as he led Company I of the Fourth Cavalry into Wawona on the afternoon of May 17, 1891. The column of soldiers first stopped in front of the opulent Wawona Hotel, where Wood, sitting ramrod straight on his horse, introduced himself to the guests on the hotel's broad porch. The Fourth Cavalry then marched farther upriver to the very spot along the South Fork of the Merced River that Newsham had recommended in his report the previous fall. Sharp orders were given as men dismounted and quickly went about erecting a row of canvas tents among the ponderosa pines.[10]

With the camp established, Wood turned to the daunting task of managing the park. He had no instructions on how to proceed, nor even a printed copy of the park rules—a fact that often put the seasoned officer into a rather foul mood. Wood knew he needed to protect Yosemite's natural resources, establish the rule of law, gain the locals' respect, or at least their acceptance of his authority, and develop a command structure that he could eventually pass on to the next superintendent. But the lack of instruction from Washington DC infuriated the captain, who would spend much of the summer pleading for further guidance on managing the park. Upon his arrival in May, he had written Secretary Noble requesting several copies of the Yosemite Act, a list of grazing regulations, and any additional suggestions for managing the park. By June, however, he still had not received a response from Noble.

Regrettably, the secretary had been away from Washington when the captain's letter had arrived, and it was buried under a backlog of delayed correspondence on Noble's desk. Nevertheless, the secretary did eventually respond to Wood's request for copies of the Yosemite Act. However, it appears that Noble's reply did not reach Wood before the captain, having grown impatient, sent another, tersely worded, letter to Washington. In his missive, Wood barked, "Up to the 30th of June, I had received no instructions whatever from the Department." Moreover, he needed clear orders on how to address the "many questions affecting the varied interests invested in this Park."[11]

It is no surprise that such correspondence crossed each other while traveling across the continent. The federal government's reliance on this method of communication often left local authorities, like Wood, to decide federal policy with minimal guidance. This reality, coupled with Noble's apparent reluctance to address the contentious issues surrounding grazing and timber interests in the park, meant that Wood determined how to manage the nation's newest national park on his own.

Without clear instructions, Wood turned to his military training and ordered a reconnaissance of the park. He lacked a detailed topographic knowledge of the park, possessing only a single copy of an 1883 survey map of the Sierra Nevada Range by U.S. Army Engineer Captain George M. Wheeler and a souvenir map he had purchased in Wawona. Neither map provided the level of detail required to administer Yosemite effectively, a point Wood emphasized in his first report to Noble. Mapping the rugged terrain proved time-consuming and challenging. At times, patrols spent weeks bivouacking throughout the park, but gradually, the cavalry became familiar with the landscape.[12]

The lack of a clear understanding of the park's terrain was just one of the pressing challenges facing Captain Wood. As park superintendent, he quickly learned that managing a national park meant managing people. This realization became particularly evident when dealing with the communities surrounding the new park. Many locals resented the park, viewing it as an encroachment on their rights to the Sierra's resources. The editors of the *Mariposa Gazette* were so incensed by the park's creation that they called for Congress to repeal the Yosemite Act on the paper's front page, arguing that "men who have no practical knowledge of this country or its resources" had locked away "untold millions" in valuable minerals in establishing the park.[13]

Moreover, landowners complained that the Yosemite Act illegally restricted access to their patented mining and timber claims and that Congress had failed to provide an opportunity for affected owners to testify before the park was established. Inholders argued that the legislation arbitrarily followed township borders instead of using the traditional metes and bounds system, which would have avoided private claims by following natural features like watersheds. After surveying the park, Wood agreed with these assertions, noting, "The private land interests within the boundaries of this park are probably much greater than Congress knew of when so much area was included within its limits." The issue of inholdings, along with pressure to open up the sugar pine forests on the park's western boundary for logging, continued to cause strife among local communities and would gradually become one of the early challenges facing the park's legitimacy.[14]

Trespassers were a persistent problem. Herds of sheep and cattle regularly encroached on the park's boundaries, damaging its foliage and trampling its meadows. Sheep were particularly problematic. Wood estimated that on his arrival there were "at least" 90,000 sheep grazing within the park. Compounding the issue, many of the herders were employed by owners who lived in the San Joaquin Valley or farther away on the coast.

Cavalry patrols spent a considerable amount of time tracking the movements of sheepherders, who were often more familiar with the terrain than the troopers, in a seemingly endless game of cat and mouse. "All of them knew just as well as we did where the boundary lines were located," Wood reported in August 1891, "but I did not wish to take severe measures until a repeated or willful trespass was committed." Adding to the challenge, when cavalry patrols did catch a trespassing flock, their options for enforcing Yosemite's boundaries were limited. The cavalry had no authority to arrest or fine violators, so they resorted to expelling the flocks and their herders to opposite sides of the park as a means of upholding the park's boundaries.[15]

By 1891 roughly thirty-five Miwok remained in the national park. In his final report on the company's first season in Yosemite, Wood noted that many of them earned a living by fishing, chopping wood, making hay, washing, and working for the valley's scattered hotels. While most continued to live in traditional cedar-built kotchas, they quickly adopted American cookware and clothing, leading many visitors, like Wood, to believe that they would eventually assimilate fully into American culture. Miwok women found a niche selling handmade baskets to visitors, which further added to their families' incomes. Such shifts had begun decades before the cavalry's arrival.

In 1872 British photographer Eadweard Muybridge, along with painter Albert Bierstadt, spent a great deal of time photographing and painting the valley's Miwok residents. Muybridge's images dispelled any anachronistic depictions of the valley's Native inhabitants, capturing candid moments within the artist's camp. To the artists, the Miwok were a curiosity as compelling as the valley's waterfalls, but not many tourists shared such opinions of the valley's Native inhabitants.[16]

California officials largely ignored the remaining Miwok, provided they remained unobtrusive. But not all appreciated their presence within the Yosemite Valley. On her encounter with a Native mother and her baby, writer and social activist Helen Hunt Jackson described the woman as half-naked and dirty beyond words, "her stiff, vicious looking hair falling around her forehead like fringed eaves, her soulless eyes darting quick glances right and left, in search of possible charity." Such a person had no place within the Yosemite Valley's sublimity, Hunt argued. Others agreed, including Muir, who later wrote that the Miwok and Paiute somehow "seemed to have no right place in the landscape." While neither Jackson nor Muir spoke for most early visitors, by the 1890s patience with the Miwok's presence was waning.[17]

Wood responded to such demands by writing in 1892 that the Native Miwok had a "moral right" to remain, though he did temper those remarks by noting that the Yosemite Indians would inevitably die off or assimilate, echoing sentiments in accord with official federal Indian policy represented in the 1890 Dawes Act. Even as the debate over the continued Indigenous presence within both the park and the grant continued into the twentieth century, the Ahwahnechee descendants continued to live within the valley, adapting to the growing restrictions on their hunting by both finding work in the nascent tourist economy and, like many settlers, clandestinely hunting away from the calvary's view.[18]

Wood led the Fourth Cavalry out of Yosemite National Park that October. As superintendent, he had set the precedent for how the park would be managed going forward and, perhaps more importantly, defined the very concept of what a national park should be. By protecting Yosemite's flora and fauna for visitors' enjoyment, Wood reinforced the idea that the park was primarily for recreation—a policy he would revisit the following year when he returned for his second tour as superintendent.

However, that spring Wood began experiencing severe pain in his mouth. By the time he led the Fourth Cavalry back to Yosemite the next summer, it was evident to all those who knew him that the grizzled veteran was suffering greatly. He soon returned to San Francisco to see a doctor and was diagnosed with tongue cancer. Ever the soldier, Wood returned to Yosemite and completed the summer's campaign despite his condition. He survived the winter but at great personal cost, undergoing several surgeries to remove large portions of his tongue. Despite these efforts, the cancer ultimately prevailed, and Yosemite's first superintendent died in April of 1894.

Wood's career spanned from the beginning of the Civil War through the nation's conquest of the Southwest and culminated in the establishment of Yosemite National Park. Throughout his time as a soldier, he embodied American power and authority, whether preserving the Union, advancing the nation's imperial ambitions, or enforcing federal authority within Yosemite National Park. It was in this last role that Wood highlighted the philosophical shift in the purpose of federal lands and the government's role in managing them. As Yosemite's first superintendent, he set a precedent for the park's purpose that would guide future generations.

AS WOOD LED his troops back into the park for a second summer in 1892, a group of men gathered in Warren Onley's San Francisco law office to discuss the formation of an outing club. Among those present was John Muir, the thin, bearded orchard farmer and author who had spearheaded the successful effort to establish Yosemite National Park just two years earlier. Muir was now deeply worried about protecting the park from grazing and timber interests. Others who joined him with such concerns included future Oakland Mayor Warren Onley; University of California professors Willis Linn Jepson, Joseph Le Conte, and William Armes; and Stanford University President David Starr Jordan. Each had received a brief invitation to the meeting three days beforehand. The idea for such a club had been circulating for several years. Congressman and publisher Robert Underwood Johnson had previously encouraged Muir to form an organization to advocate for the creation of Yosemite National Park, while a group of University of California professors had proposed creating an alpine club, similar to the Appalachian Mountain Club, to promote the exploration of California's mountains.[19]

It is uncertain which specific purpose drew each man to the office on Sansome Street that morning, but their shared passion for the Sierra Nevada certainly united them. After discussing the idea for several hours, the group decided on the name the Sierra Club and drafted articles of incorporation. The simple document encapsulated the group's recreational and conservation aims. First, the Club advocated making California's mountains accessible for members' exploration and enjoyment. Second, it sought to publish "authentic information" about these mountains for its members. Third, the Club was dedicated to "enlisting the support and cooperation of people and the government in preserving the forests and other natural features of the Sierra Nevada Mountains." As its charter expressed, the Sierra Club was as much a political organization as a mountaineering club, reflecting its founders' seemingly paradoxical desires.[20]

Club membership grew to 283 by the following winter. Predominantly middle-class, members included university professors, scientists, politicians, and local business owners. In many ways, the Sierra Club reflected the growing Progressive Movement in American politics, which saw addressing political corruption as crucial in solving the nation's social and economic ills. Many within the Club believed the Yosemite Commission was a prime example of such corruption. The Commission had long faced criticism for its nepotism and ineptitude. Outraged over the continued despoilation of the Yosemite Valley by commercial interests, members of the Sierra Club demanded the Yosemite Commission be replaced, if not removed, from managing the valley and grove.

Making matters worse, the continued trespassing of sheep and cattle into the national park threatened the very scenic values Club members sought to protect for their own recreational enjoyment. Despite the cavalry's best efforts, sheepherders and cattlemen continued to push their livestock into the High Sierra, devastating the region's delicate vegetation. In testimony before Congress, Secretary Noble had acknowledged that such "acts of despoilation" had been allowed for years and seemed to have become well-established behaviors in the region.[21]

Inholdings further fueled the controversy over grazing within the park. Under the Yosemite Act, private property owners could request permission from the park superintendent to move their stock through the park to their property within its boundaries. Park advocates viewed such drives as inconsistent with the national park's recreational purpose. Conflicts over grazing seemingly peaked in 1905 when then–Acting Park Superintendent Captain Harry C. Benson denied California State Senator J. B. Curtin the right to move his cattle on public roadways through the park. Enraged, Curtin sued, and in 1911 the U.S. Supreme Court ruled in favor of the state legislator, granting him access to his inholding. The Court's decision outraged the park's advocates, who argued that the Yosemite Act prohibited the "herding or grazing of loose stock or cattle of any kind on the government lands in the park." But despite Curtin's legal victory, the issue of inholdings' legal status remained largely unresolved.[22]

The condition of Yosemite Valley remained the primary concern among park advocates. In the same year as the Sierra Club's founding, Newsham reported to Secretary Noble that the valley resembled a poorly managed cattle ranch, attributing the poor state of the valley to the Yosemite Commission's lack of effective management. Other critics concurred, lambasting the state government for failing to adequately fund the park and for permitting concessionaires to cultivate the valley floor, create fenced-in pastures, fell trees, and dam Mirror Lake—all in efforts to increase profits by making the valley more visitor-friendly. The state commission's failures in protecting the valley from degradation were so egregious that, after an inspection, the General Land Office recommended that the state return the grant to the federal government.[23]

The idea of receding the Yosemite Grant was not new. Critics of the Yosemite Commission had long advocated returning the Yosemite Valley and Mariposa Grove to the federal government. Yet the issue gained significant attention following the establishment of Yosemite National Park. While some supported retaining the grant, arguing that its return would damage California's reputation, in truth few Californians paid much attention to the issue.

The debate over the justification for receding the grant also played out within the Sierra Club. Many members were hesitant to become entangled in a dispute over the valley's ownership, often believing that legislation would eventually resolve the issues. A few supported the Commission, while a growing number openly advocated for recession. Club member and artist Charles Robinson was particularly frustrated by the lack of the Club's full-throated support for returning the grant to federal control. In a letter to *Century* magazine's editor-in-chief, Robert Underwood Johnson, Robinson raged at the Club's apparent apathy toward the state of Yosemite Valley, accusing members of focusing more on their enjoyment of the Sierra Nevada than on their political engagement. A self-professed curmudgeon, Robinson grew increasingly exasperated by the Sierra Club's silence on the Commission's ongoing misconduct, denouncing the Club in another letter to Johnson following its first winter meeting, bellowing that despite the eulogies delivered by fellow members, "not one word was uttered condemnatory of the conditions, let alone the management of the Park's crown jewel, the Yosemite Valley."[24]

The differences between Robinson and other members underscored a growing rift within the Sierra Club regarding its mission. As the Sierra Club's president, Muir was sympathetic to Robinson's criticisms. However, other members feared that supporting the recession of the Yosemite Valley would tarnish the Club's popularity. Meanwhile, as debates over recession and grazing access continued, the cavalry sought to put an end to hunting within the park.[25]

HUNTING WITHIN THE national park had been prohibited for the past five years, a rule explicitly outlined by Secretary Noble in his original regulations. Furthermore, that previous summer, Yosemite's superintendent, Captain Alex Rodgers, had proposed a ban on all firearms within the park to eliminate hunting. His successor, Lieutenant Colonel S. B. M. Young, took the suggestion seriously, implementing a policy requiring all visitors to check their weapons at the park headquarters in Wawona before entering the park. The new rule countered decades of visitors carrying their rifles and pistols into the park under the pretext of self-defense against wildlife. Visitors had frequently used this allowance as a license to hunt wildlife for sport, which in turn had severely depleted Yosemite's deer and bear populations.

It was not surprising, then, that a hunting party entered the park in the summer of 1896. Among them were Sierra Club member Donald Campbell and prominent San Francisco businessman John Howard. Both men were well connected within San Francisco's business and political circles and had a long history of hunting in the High Sierra. It is unclear whether they were aware of the firearm ban, but their attempts to evade the cavalry patrol sent after them suggested that they not only knew of the ban but also felt it did not apply to them.

The troopers quickly overtook the hunting party and escorted them back to cavalry headquarters, where the colonel lectured them on their violations before releasing them without a fine. Incensed at their treatment, six members sent a forty-six-page complaint to both the secretary of war and the secretary of the interior on their return home, demanding an apology. Senator George Perkins, the former governor of California, was equally outraged by the group's removal from the park and sent a letter to both cabinet members calling Lieutenant Colonel Young's treatment of the party "reprehensible." The War Department responded that the cavalry was merely enforcing the Department of the Interior's rules, and Secretary Noble publicly supported Young's actions in implementing the regulation, effectively ending the matter.[26]

The arrest and removal of the six hunters cemented federal control and established the standards for appropriate visitor behavior within the national park. Noble's endorsement also reinforced the cavalry's authority to oversee and manage the park, emphasizing that the cavalry's primary purpose was to protect its natural beauty for visitors' enjoyment. However, this enjoyment was now restricted to a limited number of activities. Hunting, once a favored pastime among the wealthy and a necessary source of food and income for many rural residents, was no longer acceptable within the national park as it threatened the very wildlife that drew visitors to the park. While debates over acceptable recreational use in Yosemite continued to be controversial throughout the century, the course was firmly set by the end of Yosemite National Park's first decade.[27]

LIKE MANY TOURISTS, Caroline Van der Burg had traveled a long way to see the Yosemite Valley. Having toured much of Europe and Africa, she was well accustomed to uncomfortable travel. But as her stagecoach began its descent down into the valley its breaking blocks failed. The coach quickly began barreling down the trail, careening around corners and barely keeping from falling into the abyss below. Fearing for their lives, the passengers braced themselves as the carriage sped toward a steep gulch and corner. Miraculously, the coach swerved around the corner before crashing on its side.

After climbing out, Van der Burg and the other passengers found the driver lying a short distance away. Aside from a couple of broken ribs, he was largely unharmed but unable to drive. As the shock wore off, the group busied themselves gathering scattered luggage, righted the overturned coach, collected the horses, and settled in for the night. The next morning, a geologist, despite having little experience with horse-drawn teams, took the reins and guided the battered coach and its passengers onward to Coulterville.

Van der Burg survived the rest of the trip to the Yosemite Valley with no further incident. Over the coming days she rode to Glacier Point, sat alongside the Merced River, and awoke early to watch the sunrise over Mirror Lake, before safely continuing her Western tour. Like Van der Burg, many tourists had long described the trip to the Yosemite Valley as dreadful, leaving them bruised and covered in a thick layer of dust. The trip often proved to be dangerous as well. Only three stage roads entered the valley, one from the south and two from the north. Poorly maintained, narrow, and steep, none was trouble-free. Particularly on the descent into the valley, in which stagecoaches pulled by teams of horses had to navigate several sharp switchbacks before arriving on the valley's floor. Such difficulties kept many from visiting the national park and Yosemite Valley until the Southern Pacific Railroad's completion of a spur line to El Portal in 1907.[28]

Travel to the park remained challenging despite the railroad, which primarily hauled timber. The majority of tourists continued to make their way to the valley via the three stage routes until the 1910s, when, for a brief time, the Yosemite Railroad opened the park to a growing number of tourists. Because of its relatively late entry into the Sierra Nevada, the railroad was soon overrun by America's newest obsession, the automobile.

CROWDS QUICKLY GATHERED on the front porch of the Wawona Hotel as Oliver Lippincott and Edward Russell arrived in their steam-powered Locomobile. Manufactured in Connecticut, the automobile was so compact that it barely accommodated both men as it chugged to a stop. Weighing roughly three hundred pounds, Lippincott occupied most of the Locomobile's single bench seat, leaving only a tight space for Russell to sit. The two men had brought the Locomobile from Los Angeles to Fresno via the Southern Pacific Railway, from where they then drove to the small farming community of Raymond before continuing on to Wawona. Lippincott hoped the trip would generate publicity for his future photography studio in the Yosemite Valley. Meanwhile, Russell, a machine shop owner, accompanied him principally in case the temperamental Locomobile broke down. The Locomobile itself was a relatively simple machine—an open carriage mounted on four wooden wheels, with a single-piston steam engine positioned behind the narrow upholstered bench seat. Upon seeing the vehicle, one local remarked that it looked like two bicycles with a seat in the middle.

The following morning, the two adventurers made the thirty-mile drive from Wawona to the Yosemite Valley in a remarkable three hours. They then spent the next several weeks touring the valley, with Lippincott staging several photographs of himself and Russell racing around in the Locomobile. The most memorable image featured Lippincott and a guide standing in the vehicle, secured by ropes around their waists, on the overhanging Glacier Point. "I firmly believe that if the machine had gone over, every man of the party would have gone with it," Lippincott later wrote of the escapade.[29]

News of Lippincott and Russell's trip to Yosemite quickly spread, and within five years, automobiles became a common sight within the park.

Not everyone welcomed the transportation revolution, however. Owner of the Yosemite Stage and Turnpike Company Edward Washburn ardently believed that automobiles threatened the safety of both his passengers and horses. Citing Yellowstone's exclusion of automobiles, he wrote to Yosemite's superintendent, Major Harry C. Benson, requesting a similar ban within Yosemite. Benson shared many of his concerns about the increasing number of automobiles and the condition of the park's roads and so was sympathetic to Washburn's request.

Rough even under the best conditions, Yosemite's roads were challenging for horse-drawn stages to navigate, let alone for automobiles prone to breaking down. After experiencing numerous breakdowns while attempting to drive to Yosemite Valley, one visitor bitterly remarked, "The roads are poor, and the grades are almost insurmountable." Furthermore, automobiles required fuel, water for their radiators, and spare parts, all of which had to be transported into the valley by stage. Considering these issues and the precedent set by Yellowstone's superintendent, Benson ordered all automobiles banned from entering Yosemite National Park in the summer of 1907.[30]

Rather than settling the issue of automobiles in Yosemite, Benson's decree only intensified calls for their admission into the park. Within two years of the ban, the Sierra Club's Board of Directors wrote to Secretary of the Interior Richard Ballinger, advocating for the allowance of automobiles. Recognizing the conflict between horse-drawn stages and automobiles, the board stated, "We feel that the automobile, as a means of travel, has become a permanent part of our life, and sooner or later, its admission to the Park will certainly follow." The board argued that it would be more effective for the government to regulate where and when automobiles could travel within Yosemite. Despite these protests, Interior Secretary Richard Ballinger maintained that the park's roads were inadequate for automobile travel, and that lifting the ban would be "tantamount to eliminating travel by stage."[31]

The prohibition lasted just a few short years before the Department of the Interior yielded to the nation's growing embrace of the automobile. In 1913 the Interior Department lifted the automobile ban. Improved roads followed over the next two decades, transforming Yosemite into a windshield wilderness. While automobile proponents celebrated the decision, some visitors, such as Great Britain's Lord James Bryce, warned, "If Adam had known what harm the serpent was going to do, he would have tried to prevent it from finding lodgment in Eden; and if you were to realize what the result of the automobile will be in that incomparable valley, you would keep it out." Not everyone shared Bryce's view. Most famously, John Muir welcomed automobiles into the park, describing them in a letter to a friend as "useful, progressive, blunt-nosed mechanical beetles." Opening Yosemite to a greater number of visitors, Muir reasoned, would only help expand Yosemite's popularity and deepen support for its continued preservation. While such reasoning had its apparent flaws, Muir's boosterism drew ever larger crowds not only to Yosemite but to the growing number of national parks.[32]

BY 1898 THE Sierra Club was in a financial bind. In his annual report, the Club's secretary, Robert Price, noted that many member dues remained unpaid, making it difficult for the Club to cover the rent on its offices and the costs of its biannual bulletin. He admonished members to square their accounts quickly, or the Club would soon face insolvency. Reasons for delinquent dues varied, but the most common was members' lack of engagement with the Club. Most active members had joined because they were drawn not by the outdoors, but by the Club's political activities. The Sierra Club's leaders lamented members' seeming apathy, fearing it would undo the Club's broader conservation goals. Yet, at that same moment, the Sierra Club was undergoing a generational shift. Younger members like San Francisco lawyer William Colby were taking on greater leadership roles within the Club, embracing both its politics and its call to bring more Californians outdoors.[33]

Inspired by Muir's writings, Colby first explored the Sierra Nevada in the spring of 1894. For three months the nineteen-year-old student joined two classmates hiking across the High Sierra, first traversing the Hetch Hetchy Valley, then summitting several peaks before making the arduous trip down the Tuolumne Canyon. The three friends had spent weeks planning the trip while hiking Mount Grizzly outside Berkeley, often pausing on its summit to gaze up at the stars and revel at the possibility of adventure in the High Sierra. The trip opened Colby's eyes to the mountain range's raw beauty. "I realized that travel of that sort was the very highest order, something that I repeated largely during the rest of my life as long as I was able to go into the mountains and do the strenuous hiking that was necessary for such trips," he later told an interviewer. Colby would spend the rest of his life doing just that.[34]

Over the next four years, Colby dropped out of school, unable to afford tuition, and took on teaching to make ends meet. Then he finally returned to school and earned his law degree, from Hastings College of Law, specializing in mining and forestry law. But unlike his classmates, Colby did not immediately seek a clerkship following graduation. Instead, he took a summer job assisting visitors in the Sierra Club's public reading room and visitor center in the Yosemite Valley.

It was a heady time to be in the Yosemite Valley. America's war with Spain meant the soldiers who had patrolled Yosemite's backcountry for the previous seven summers were now absent. In their stead, the federal government hired Archie Leonard and Charlie Leidig as the park's first civilian rangers. That summer, workers refurbished the Sentinel Bridge, and John and Bridgett Degnan opened their new bakery, selling nearly a hundred loaves of bread a day. The new bakery was the start of a boom of new businesses within the valley. Six new art studios and Camp Curry opened over the next six years, each seeking to capitalize on Yosemite's growing popularity as a tourist destination. When not helping visitors, Colby spent the summer exploring the park before returning to San Francisco in the fall to clerk at a local law firm. But Yosemite was never far from his mind. Even as he opened his own practice specializing in mining law, Colby continued to explore the park as often as he could.[35]

Colby was elected as the Sierra Club's secretary in 1900, replacing Price. That summer, Muir invited the young lawyer to his home in the Alhambra Valley. The two men spent several days walking through Muir's orchards discussing the challenges facing the Sierra Club. They agreed that one of the most pressing issues was garnering wider public support for the recession of the Yosemite Valley. Muir had long been frustrated with Californians' growing apathy toward nature, and he wondered aloud if there was some way to draw people into the Sierras to experience them as both he and Colby had. Colby concurred that if people experienced the Sierra Nevada, then they too would form the same connection to these wonderous places, and in doing so they might like to join the fight to save the Sierras from logging, mining, and further commercial development.[36]

It was an inspired idea. Colby jumped on it immediately, proposing the Sierra Club invite its members on a month-long outing in Tuolumne Meadows the following summer. Nearly one hundred made the inaugural trip. Even more joined the next summer's outing to Kings Canyon. Within two years, Colby joyfully reported the Club was debt free. More importantly, he noted, the new members embraced the Sierra Club's preservationist efforts, "because they are in sympathy with the Club's work and appreciate the advantage they receive from such membership." Looking ahead, he foresaw similar trips could include Mount Rainier, Yellowstone, and even Alaska.[37]

The same year, Muir published *Our National Parks*. Composed largely of his earlier essays on the national parks, the book laid out his argument for their preservation. Muir opened with an essay he had first published in the *Atlantic* in 1898. In it, he proclaimed, "Thousands of tired, nerve-shaken, over-civilized people are beginning to find out that going to the mountains is going home; that wilderness is a necessity; and that the mountain parks and reservations are useful not only as fountains of timber and irrigating rivers, but fountains of life." It was a telling statement that emphasized recreation's true value within the national parks.

Muir then went on to describe all three parks, including Yosemite's forests, wildlife, birds, and streams, before calling on the federal government to save Yosemite from those who sought to destroy this wilderness temple. "God has cared for these trees, saved them from drought disease, avalanches and a thousand straining, leveling tempests and floods," Muir wrote, "but he cannot save them from fools—only Uncle Sam can do that." Within two years he would be able to make that very case to a very influential man from Washington.[38]

THE THREE WOMEN watched expectantly down the narrow dirt path. They wore warmer dresses for the early spring weather along with large, fashionable hats that shielded them from the Sierra sun. It had snowed the previous night, so a chill still tinged the morning air. The Merced, swollen with spring snowmelt, roared beneath the Sentinel Bridge, where they stood. And yet, a feeling of anticipation filled them as they waited. Then a group of five men on horseback appeared in the distance. A stout man dressed in a field jacket, loose-fitting jodhpurs, and a wide-brimmed hat rode in front. Sunlight glimmered off his spectacles, emphasizing his mustache. The women smiled and rushed to the rider, hoping to have a moment to shake his hand. For it was not every day they could meet the president of the United States.

Seeing the three women standing at the bridge ahead, President Theodore Roosevelt slowed his horse. Glancing back, he spoke briefly to his guide, Charlie Leidig, who after a quick nod spurred his horse forward. Just as the group approached the bridge, Leidig reared his horse, startling the women, forcing them back, and allowing the president to speed by the three would-be well-wishers. Standing on the other side of the bridge, a visitor captured the moment in a hastily shot photograph. The picture is blurry, caught in a moment of excitement and perhaps chance. The candid shot captured the president glancing down at the photographer as he hurried past the women.[39]

Having navigated the crowded bridge, the president's party pulled up in front of Angela and Chris Jorgensen's stout log home and art studio. There, a small group welcomed Roosevelt, who quickly shook everyone's hand before entering the studio. Minutes later, after just briefly admiring Jorgensen's watercolor paintings, the president reemerged and thanked the couple but regretfully declined their hospitality, explaining he planned on spending one more night under the stars before leaving Yosemite the next morning. With that, Roosevelt remounted his horse and rode off with Muir and his fellow companions. The Jorgensens were saddened at the president's fleeting visit; plans had been made for a full evening of dining, speeches, and fireworks. The crowd milled about in front of the studio deciding what to do next. Some attempted to follow the president, while others simply wandered off to enjoy the remainder of the day.

Roosevelt was finishing a two-month tour of the American West when he arrived in Yosemite that spring. He had traveled through North and South Dakota and spent two weeks roughing it in Yellowstone with naturalist John Burroughs before crisscrossing the region from Omaha to Los Angeles and to the South Rim of the Grand Canyon. When asked what should be done with the Grand Canyon, he had answered, "Leave it as it is. You Cannot improve on it." Months before, Roosevelt had written Muir asking him to join him in Yosemite. Muir readily agreed, believing he could "do some forest good in talking freely around the campfire" with the president.[40]

No president before, or since, has been so enthusiastic or so informed about the nation's wild places as Roosevelt. Likewise, Muir had become a national celebrity known for his writings calling for the preservation of wild places. Both men personified America's changing relationship with nature, a relationship that sought to conserve the nation's natural bounty. However, the purpose of conservation remained a matter of debate at the turn of the century.[41]

Muir joined the president's entourage in Berkeley, from where they traveled by train to the farming town of Raymond. A crowd awaited the president at the small station. Perturbed at the delay, Roosevelt gave a brief speech before loading into a passenger coach with Muir. Then he was escorted by a squad of Ninth Cavalry "Buffalo Soldiers" assigned to the park for the summer in the Mariposa Grove. There, other dignitaries joined, including California Governor George Pardee, for a photograph at the base of the Giant Grizzly. After the photograph, Roosevelt quickly dismissed all but Muir, guides Charlie Leidig and Archie Leonard, and an army packer named

Jacker Alder for one of the most memorable, and consequential, visits in national park history.

Having traveled thousands of miles, the president was exhausted by the time he reached Yosemite and was in no mood for crowds. After separating from the rest of the Presidential entourage, Roosevelt pulled Leidig aside and asked the guide to steer the group away from the crowds so that he could get some rest and enjoy his time in the park. Leidig noted he would and set about making a meal of fried chicken and beef steak, which the president washed down with cups of strong black coffee before retiring for the evening in the shelter of the Grizzly Giant. Over the subsequent three days, the five men slept outdoors, posed for photographs at Glacier Point, ate their meals over campfires, and, as Leidig later reported, talked incessantly about birds, trees, Muir's glacial theory on the formation of the Yosemite Valley, and most importantly the future of conservation.

The men broke camp early the next morning. Roosevelt repeated his desire to stay away from "civilization," to which the rest of the group readily agreed. Rather than heading directly to the Yosemite Valley, they made their way to Empire Meadows, and after eating lunch they decided to ride to Sentinel Dome. The troupe plowed through deep snow along the way. At one point Leidig had to help the president out of a drift. With heavy snow still falling, the men discussed where to escape the weather for the evening. Muir suggested they camp on the back side of the Sentinel Dome, but Leidig insisted they camp at Glacier Point. Muir eventually agreed.

Awakening to nearly half a foot of snow, the five swiftly broke camp and headed to Glacier Point, where Muir and Roosevelt posed for photographs standing on the granite ledge. The task done, they mounted their horses again and rode to the Little Yosemite Valley. Much to Roosevelt's dismay, a crowd had gathered in the valley hoping to see the president. Skirting the multitude, the men made their way into the Yosemite Valley, where they encountered the three women waiting for the president at the Sentinel Bridge. After dispensing with the formal visits, the men mounted their horses and rode out to find a camp spot for the final night.

The five settled at a likely spot near Bridalveil Meadow. But a crowd soon followed. Once again annoyed at being interrupted, Roosevelt asked Leidig to get rid of the throng. Charlie obliged and walked out into the meadow to inform those seeking a glimpse that the president was very tired and wished to be left alone. Chagrined, the crowd slowly melted back to the Old Village, where fireworks celebrated Roosevelt's visit. After dinner Muir and

Roosevelt wandered out into the meadow until dark. We do not know what the president and naturalist talked about that evening as they watched the sunset. It could have been about the valley's geologic history, its birds, the pressing need for Roosevelt to designate forest reserves, or the recession of the Yosemite Valley and Mariposa Grove from the state of California. But after their talk, the two returned to the fire, where the president regaled the group with stories of lion hunting in Africa before turning in for one last night under the Sierra stars. The following morning the crowds returned. And so, after a rushed breakfast, Roosevelt and Muir shouted their goodbyes to the three other men and hurriedly boarded the stage.[42]

A wooden sign sits near the spot where Roosevelt and Muir camped that last night. Its brief inscription tells readers that Muir urged the president to preserve the "priceless remnants of America's wilderness." And in doing so, he inspired Roosevelt to do just that. Few visitors stop at the historical marker, as most whiz past it on their way to the Yosemite Village. Yet it marks one of the most important moments in the park's history. Roosevelt proved to be the friendly and powerful ally Muir sought in supporting the recession of the Yosemite Valley and Mariposa Grove.

Yet it would take another three years before a bill appeared on the president's desk placing both the valley and the grove under federal jurisdiction as part of the national park. During that time, Congress eliminated 542 square miles from the park to placate timber and mining interests and debated ceding the Hetch Hetchy Valley to San Francisco for the construction of a reservoir. In each case, Roosevelt proved to be both a proponent of the conservation ethos he and Muir had discussed, during their time roughing it through Yosemite, and a savvy politician knowing that he had to bargain with far larger forces outside of the park.

IT HAD BEEN a challenging four days hiking down the Tuolumne Canyon. Having started at the river's headwaters in the Tuolumne Meadow, the fourteen men crashed through miles of thicket, following what few game trails they could find. As they labored deeper into the canyon, they struggled scrambling over barren granite boulders while avoiding rattlesnakes. On the second evening, they feasted on freshly caught trout while camped alongside the Tuolumne as it cascaded into turbulent pools.

The vegetation changed rapidly as the men made their way down the valley, and the water grew warmer, making it more difficult to fish. Poison oak seemed to lie everywhere, making navigating downstream that much more difficult. Magnificent California black oaks dotted the valley floor, as the Tuolumne River languidly flowed by. In the growing twilight, the fourteen finally heard the revelry of their fellow Sierra Club members. Relieved at finding the main group, they enjoyed a quick dinner of their remaining provisions and fell asleep under the branches of an enormous oak, which Club member William Frederic Badè later mused had probably never spread its boughs over such a party.[43]

The Sierra Club's 1905 outing was most likely the largest group of tourists to visit the Hetch Hetchy Valley to that point. The valley was in a remote and largely unknown corner of the park. For centuries, local Paiute and Miwok nations had enjoyed its relatively mild weather, plentiful fish, and beauty. But since the cession of the Yosemite Valley, few tourists had bothered to venture into Hetch Hetchy. So few, in fact, that the first campground was not developed in the valley until 1904. Yet those who did make their way to the valley proclaimed its beauty rivaled that of the Yosemite Valley.[44]

Two Chinese cooks were among those on the outing. Largely absent in the Club's retelling of the outing in its bulletin, the two undoubtedly remained in camp—while the others explored the area—preparing the meals and setting up and cleaning the cook tent. Chinese cooks were common throughout the Sierra Nevada, running not only camp kitchens but also many of the region's hotel kitchens. A few gained renown for their culinary skills, including Tie Sing, who famously cooked for both of Park Service Director Stephen Mather's Yosemite trips in 1915 and 1916, during which the national park director brought several dignitaries into Yosemite's backcountry to convince them to support the nascent National Park Service. Tie Sing and his assistant, Eugene, arose every morning before dawn to start the fire, cook a hearty breakfast, and prepare lunches. After serving the meals, they cleaned the dishes and linens before packing them onto a mule and hiking to the next campsite. So magical were the meals prepared by these two men that Mather often credited them with helping to establish the Park Service.[45]

The outing's two cooks' routine most likely mirrored that of Tie Sing and Eugene a decade later. However, their invisibility in the Club's account of the two weeks is telling. Chinese laborers, whether as cooks, road builders, launderers, or loggers, played a pivotal role in the development of the Sierra Nevada. Early immigrants worked in the region's gold mines, helped build both the Wawona and Tioga Roads, cut cordwood, and, most famously, worked for the Central Pacific Railroad, blasting the transcontinental railroad from California to Utah. While often regarded as dedicated and hard workers, Chinese immigrants faced unrelenting racism that often erased their contributions. Tie Sing was an exception. In 1899 the U.S. Geological Survey named the 10,540-foot Sing Peak within Yosemite National Park in honor of his contributions to surveying the Sierra Nevada. Most Chinese immigrants, however, did not receive such recognition, and their experiences within Yosemite have largely been lost in the history books.[46]

Perhaps most importantly, the 1905 outing also marked the opening of the Club's campaign opposing San Francisco's plan to impound the Hetch Hetchy Valley, and a political battle that would shape both Yosemite and the future of American environmental politics.

FOR A HALF a century, the city of San Francisco was the center of power in California. Wealth extracted from the Sierra Nevada's streams and hillsides passed through the port city, turning it into the largest metropolitan area on the Pacific coast. Its growth necessitated water, something in short supply in the Bay Area. So much so, that by the late 1890s the city's mayor, James Phelan, cast his eye east toward the Sierra Nevada. There, tucked within the newly created national park, lay the Hetch Hetchy Valley. The fight over the valley raised a question: What was its highest and best use? Was it providing abundant clean water to the largest city in California? Or was it the protection of a national park? The answer reflected both California's political reality at the beginning of the twentieth century and the beginning of what would become the modern environmental movement.[47]

Early attempts to provide water in the San Francisco Bay Area had met with limited results. Private waterworks firms built an expansive, albeit modest, system of earthen dams and flumes to draw water from across the peninsula to San Francisco. The region's frequent droughts, combined with political struggles over the city's water supplies, proved too much to overcome, and as the nineteenth century drew to a close it became readily apparent to all that private interests could not provide enough water to quench the thirst of the city's burgeoning population. The solution to San Francisco's water issues, many came to believe, lay in the far Sierras.

City engineers began eyeing the Tuolumne River as a possible solution to the city's water needs as early as the summer of 1892. Draining a large expanse of the Sierra, the river slowed as it entered the Hetch Hetchy Valley, meandering languidly through it before plummeting through a narrow V-shaped gap on the valley's western end. From there, the Tuolumne flowed into the Central Valley, joining the San Joaquin River, which in turn flowed into the San Francisco Bay. City engineers fell in love with the valley's narrow western outlet, noting its shape and geology made it as nearly a perfect dam site as could be found anywhere within the Sierras. The fact that the Hetch Hetchy Valley lay wholly within a newly established national park did little to dissuade the city's mayor, James Phelan. Born and raised in the Bay Area, the successful lawyer and banker had become mayor in 1897. Understanding that the only way the city would continue to grow and thrive would be through acquiring a sufficient and stable water supply, Phelan filed for water rights within the Hetch Hetchy Valley in 1901.

Despite the city's need, Interior Secretary Ethan Hitchcock rejected Phelan's application outright, citing that as secretary he did not have the authority to grant a right of way over private lands, through which much of the proposed aqueduct would pass. Furthermore, the 1890 Yosemite Act stipulated his office preserve and retain the park in its "natural condition." Frustrated by Hitchcock's decision, Phelan refused to give up on his plan. Likewise, Muir and the Sierra Club steadfastly opposed the reservoir, seeing it as akin to damming the people's cathedrals and churches as water tanks.[48]

The thought of flooding the valley appalled Sierra Club members, who viewed it as a violation of the national park ideal. And so the 1905 summer outing was as much a political gathering as it was an exploration of Yosemite's backcountry. By visiting the Tuolumne River, Club members sought to demonstrate the valley's worth for leisure and create a deeper sense of place among its members. It was a proven strategy that had worked in convincing President Roosevelt to further the Sierra Club's efforts in receding the Yosemite Valley just years prior, and the Sierra Club hoped that it would once again work to save the Hetch Hetchy Valley.

Fate struck the following year when an earthquake devastated San Francisco. An ensuing fire ravaged much of the city, leaving it in ruins and making many of its citizens refugees. The tragedy roused national sympathy for the city, providing Phelan the political capital he needed to secure his reservoir. The fate of the Hetch Hetchy Valley became a national issue, pitting competing visions of progressive reforms against one another. Phelan believed that providing San Francisco with a reliable and safe source of water was the highest public calling. Muir and members of the Sierra Club countered that the valley's preservation would better serve the public's interests.

The political battle continued following the 1908 presidential election. Seeking to win newly elected President William Howard Taft's support, the Sierra Club invited the new president to Yosemite to experience the park himself. Taft agreed, and in the following summer he traveled to the park with Secretary of Interior Richard Ballinger. With Muir as a guide, the two toured the Mariposa Grove and Yosemite Valley. The Club's lobbying efforts worked, and Ballinger once again rejected the city's proposal to build a dam within the national park the following year. But the Hetch Hetchy's future was thrown into doubt again after Ballinger was forced to resign, due to a series of controversies, soon after making his decision on the valley. A report by the Army Corps of Engineers the following year supported the view that the Upper Tuolumne River watershed provided the best alternative for San Francisco's drinking water. Fearing further political backlash, Secretary of the Interior Ballinger's replacement, Walter L. Fisher, deferred to Congress on a final decision on the valley's fate.

In 1913 California Representative John Raker, in whose district Yosemite lay, introduced legislation to allow San Francisco to build the water delivery system. The Raker Act, as it became known, granted the city the right to build its dam and flood the Hetch Hetchy Valley. The act included several regulations, including requiring the city to provide public access. President Woodrow Wilson signed the bill into law December 19, 1913, explaining, "I have signed this bill because it seemed to serve the pressing public needs of the region concerned better than they could be served in any other way and yet did not impair the usefulness or materially detract from the beauty of the public domain . . . I believe the bill to be, on the whole, in the public interest." It was the Sierra Club's largest political defeat, and its largest victory. In the decades to come, the Club pointed to the loss of the Hetch Hetchy Valley as a cautionary tale about the costs of development within the national parks.[49]

THE RAKER ACT's passage marked a critical turning point in an increasingly heated debate over the federal government's management of its growing number of national parks. By 1910 five new national parks, Mount Rainier, Crater Lake, Wind Cave, Mesa Verde, and Glacier, had joined Yosemite, Sequoia, and Yellowstone. The passage of the Antiquities Act in 1906 added a dozen National Monuments during the same time. Furthermore, the United States' increased military presence across the globe brought the army's continued management of national parks into question. This combination of rapid expansion and dwindling Department of War support strained the Department of the Interior's administrative ability. While the Department nominally managed the parks and monuments strewn across the American

West, a growing number of advocates saw the need for a national park bureau to manage the parks in a more businesslike manner.

Furthermore, Congress's granting the Hetch Hetchy Valley to San Francisco, and Yosemite's boundary reduction, caused many within the Sierra Club to fear for the park's future. John Muir's death in 1914 further eroded Club morale. Muir's failing health had kept the famous conservationist from testifying against the Raker Act and left the Club's leadership in disarray. Yet the legislative defeats served to galvanize the call for the creation of a national park bureau by a growing number of influential Easterners. Chief among them was the president of the American Civic Association and Pennsylvania native J. Horace McFarland.

The small river town of Harrisburg, Pennsylvania, grew into a booming steel and rail city in the decades following the Civil War. While industrialization brought tremendous wealth to the burgeoning city on the banks of the Susquehanna River, it also brought sewage, polluted air and water, and disease. A native son of Harrisburg with a successful printing business, McFarland helped the city embrace what would become known as the City Beautiful Movement by developing a new park system, paved boulevards, and most importantly a modern sewage system in efforts to modernize Harrisburg.

McFarland gained national notoriety for his efforts in Harrisburg, helping him become the president of the American Civic Association in 1904. The progressive organization promoted the ideals of the City Beautiful Movement across the nation, seeking to better American life through the preservation of nature and promotion of the arts. As the association's president, McFarland was well aware of the fight over saving the Hetch Hetchy Valley. The passage of the Raker Act enraged the progressive reformer, who saw it as a direct attack on the national park ideal. Joining Muir and William Colby, McFarland began advocating for the creation of a national park bureau to prevent future such incursions into the national parks.[50]

In 1909 McFarland approached President Howard Taft and Secretary of the Interior Richard Ballinger on supporting the creation of a national park service. The president agreed on the need to better manage the national parks, but the politics of creating such an agency were complicated. The head of the newly formed U.S. Forest Service, Gifford Pinchot, openly opposed the creation of a national park agency, arguing it would erode the Forest Service's ability to manage the nation's forests. In response to Pinchot's opposition, McFarland recruited Frederick Law Olmsted Jr. to the cause. The youngest son of the famous landscape architect, and former head of the Yosemite Commission, the younger Olmsted was an obvious choice to help create such an agency. After graduating from Harvard, he joined his brother John at the architectural firm of Burnham and McKim. Over the next several years, Olmsted Jr. worked with both Burnham and his brother John Charles Olmsted designing numerous city parks, including the redesign of the aging Washington Mall.

With a renowned name and experience on some of the nation's highest profile landscape architecture projects, Olmsted Jr. had established himself as a leading authority on the significance of urban parks. Olmsted recognized the necessity of centralizing national park management but viewed the parks more through the lens of urban design than wilderness. For him and his contemporaries, national parks represented a natural extension of the City Beautiful Movement, requiring thoughtful planning that prioritized recreational use of the outdoors. This perspective guided his drafting of the key section of the 1916 Organic Act, mandating the National Park Service "promote and regulate the use" of national parks and monuments.[51]

Yet the national parks were clearly not urban. In 1910 every national park and national monument lay in the American West, and few sat near urban centers. Understanding this political reality, Olmsted suggested that a "Western man" with the instincts of a gentleman should head any new bureau overseeing the parks. And he believed he had found his man in Stephen Mather. Born in San Francisco, Mather graduated from the University of California in 1887. After working as a reporter for the *New York Sun*, he went to work with his father at the Pacific Coast Borax Company, where he became very wealthy. In 1905 he joined the Sierra Club outing to Mount Rainer. The trip reignited his passion for nature, and he soon became an ardent mountaineer and conservationist.[52]

Yosemite played a central role in Mather's becoming the National Park Service's first director. Like many Californians Mather loved Yosemite, visiting the park as often as he could. The story went that after visiting Yosemite, he wrote to then–Secretary of the Interior Frederick Lane in 1914 complaining about the poor condition of the park. Frustrated with Mather's complaint, Lane reportedly wrote back, "If you don't like the way the parks are being run, come on down to Washington and run them yourself." It is a clever story. But Mather's assistant Horace Albright later questioned it, noting Mather and Lane did not personally know each other at the time. The more likely truth was that Mather joined McFarland, Olmsted Jr., and author Robert Sterling Yard in lobbying for the creation of a federal park agency.[53]

The faction's campaign to create a national park service promoted the national parks' recreational value as the primary reason for the bureau. Like timber and grazing, recreation provided a utilitarian, and by extension an economic, value, they asserted. Such framing reflected the era's progressive politics as well as the political leverage needed to pass the National Park Service Organic Act. Signed into law by President Woodrow Wilson in 1916, the Organic Act, as it became known, mandated that the newly created Park Service "promote and regulate" the national parks, whose fundamental purpose was to "conserve the scenery and the natural and historic objects and the wild life [*sic*] therein and to provide for the enjoyment of the same in such manner and by such means as will leave them unimpaired for the enjoyment of future generations."[54]

Generations since have debated the sentence's true meaning, arguing it supports their particular vision of the National Park Service's mission. However, its author's primary concern was the preservation of each national park's scenery for the express purpose of visitors' enjoyment. The Organic Act's mandate, then, was understood to be recreational. This interpretation would change significantly over the next fifty years, but its initial intent reflected its authors' understanding that national parks like Yosemite were for recreational use—an understanding shared by the park's ever-growing number of visitors.[55]

FIG. 1. Thomas Ayers drawing of Yosemite Valley, 1855. Yosemite Research Library.

FIG. 2. Galen Clark in the Big Tree Grove, between 1858 and 1859. Photograph by Carleton Watkins. Yosemite Research Library.

FIG. 3. Members of U.S. Fifteenth Cavalry, Troop I, on Fallen Monarch in the Mariposa Grove of Giant Sequoias, ca. 1901. Yosemite Historic Photo Collection.

FIG. 4. (*opposite top*) Wagons making their way up Big Oak Flat Road from Yosemite Valley, 1903. Photograph by Julius Theodore Boysen. Yosemite National Park Archives.

FIG. 5. (*opposite bottom*) President Theodore Roosevelt crossing Sentinel Bridge, 1903. Unknown photographer. Travel album of Yosemite, Sierra Nevada Mountains, Yellowstone, and other sites, 1903–1905. Huntington Library.

FIG. 6. (*above*) President Teddy Roosevelt and John Muir standing on a rock at Glacier Point, Yosemite, 1903. Yosemite National Park Archives.

FIG. 7. Tourists driving through the Wawona Tree, 1918. Keyston View Company, *The Wawona Tree, Mariposa Grove, Yosemite Valley, Cal. U.S.A.* Meadville PA: Keystone View Co., June 15, 1918.

FIG. 8. (*opposite top*) Traffic at the Old Yosemite Village. Undated, most likely during the 1920s. Yosemite National Park Archives.

FIG. 9. (*opposite bottom*) New Village Plaza sometime around the mid-1920s. Yosemite National Park Archives.

FIG. 10. Campers in Stoneman Meadow, 1927. Mariposa County Library.

FIG. 11. Congressional party at Hetch Hetchy, 1933. Yosemite National Park Archives.

FIG. 12. Crowd watches as Curry Company employee pushes burning coals from atop Glacier Point to start the Firefall. Photograph by Arnold Williams. Yosemite Research Library.

FIG. 13. Chris Brown with a child on his knee at 1929 Indian Field Days. Yosemite National Park Archives.

FIG. 14. Chris Brown with Maggie Howard in front of Yosemite Museum, 1931. Yosemite National Park Archives.

FIG. 15. Brown bear investigating a car in Yosemite Valley, approximately 1930. Photograph by C. C. Pierce. C. C. Pierce Collection of Photographs, Huntington Library.

PART 2

3 California's Playground

The railcar was beastly hot, but at least the views were nice, Gladys Apker silently mused as she watched the Merced River slowly roll by outside her window in the summer of 1922. She had boarded the train in Oakland that morning and traveled east through the Central Valley to the farming community of Merced. There, Gladys's passenger car had been transferred to a Yosemite Valley Railroad engine, which was now slowly pulling the train into the Sierras. Piles of mining tailings along the side of the tracks told of the region's mining past. Gladys lazily listened as the tour guide pointed out the window at a pile of bricks that had once been the first bank serving gold prospectors in the region. Half listening to the guide, she grew increasingly restless as the train trudged up the Merced Canyon. Finally, the train pulled into the El Portal station, where she and her fellow travelers disembarked, before being hurriedly loaded into nearby buses with their luggage and driven into the Yosemite Valley.

Passing by Bridalveil Fall, Gladys caught her breath at the sight of El Capitan as its towering granite face came into sight. She stared in wonder at Yosemite Falls as the bus made its way through the Old Village past the grand Sentinel Hotel, which in truth did not look all that grand these days, turned right at the Sentinel Bridge, and passed the Chinese quarters. Driving under the wooden arch that welcomed them, the bus finally pulled up in front of Camp Curry. A flurry of activity erupted as porters raced to grab the exhausted travelers' luggage, and hosts shepherded everyone off the bus and toward registration.

Gladys was famished after the arduous day's journey. But instead of finding a table in the camp's cafeteria, she took a seat next to the campfire outside. Sitting down with a young woman she had met on the train, Gladys enjoyed the next couple of hours of orchestra music, violin and soprano solos, and funny sing-a-longs. As darkness fell over the camp, the crowd hushed in anticipation. A call rang out, "Hello Gallagher!" A tiny voice from above responded, "Hello, Camp Curry!" Then the voice from camp replied, "Let 'er go Gallagher!" A stream of red embers poured off Glacier Point above, producing the celebrated Firefall. Violin music filled the air as the embers slowly winked out in the twilight. With the evening's events at a close, Gladys finally turned in for the night.

The next morning, Gladys joined a group on a hike to Happy Isles, snapping as many pictures of the valley as she could along the way. That afternoon, she boarded a bus to the Glacier Point Hotel. After checking in, she found a place to sit on the hotel's veranda overlooking Half Dome. "Sunset that night was gorgeous," Gladys later wrote her parents. She lingered, watching the moon cast the whole Sierra in silver moonlight, before heading to bed.

Gladys rose early again the next morning and hiked to the Sentinel Dome, having to race back to catch the bus to Wawona for a quick lunch and tour of the Mariposa Grove. She walked in wonder through the grove of giant sequoia, pausing at the base of the Giant Grizzly and passing through the tunnel carved through the Wawona Tree before happily discovering the Wisconsin tree, named after her home state. The next day Gladys took a bus back to Merced and boarded the train to San Francisco to spend a few days in the city. After sailing along the California coast on a cruise ship to Long Beach, Gladys finally arrived back home at nearly midnight, exhausted but having enjoyed every minute of the trip.[1]

Among the growing number of Californians visiting Yosemite, Gladys arrived during a time of significant change in the park. That same summer, only two-thirds of visitors traveled to Yosemite by automobile. By the end of the decade, almost all visitors were arriving by car. The completion of the All-Year Highway in 1926 made the park a day's drive for many Californians, especially those living in the state's rapidly expanding cities.

But it was not just about cars. Recreation itself was changing, and Yosemite was starting to feel more like the state's playground than a remote escape. By the 1920s national parks weren't just seen as peaceful, scenic getaways but increasingly as places for fun and entertainment. People still came to admire the views, of course, but they also expected more to do while they were there. The Park Service and its concessioners were all too happy to deliver, adding swimming pools, golf courses, a ski area, and even a zoo to keep visitors entertained and sticking around longer.

Such amenities came to define tourists' experiences in Yosemite, shaping the way visitors engaged with its natural landscape. Over the course of the next two decades, the park evolved from a rugged wilderness into a modern destination resort, complete with a lavish lodge that offered comfort amid the scenery. The addition of a museum provided cultural and historical context, enriching the educational aspect of the visit, while a seasonal summer festival brought a sense of community and celebration. Together, these developments transformed Yosemite into a place where nature and leisure intertwined, appealing to a broader public and leading to further development throughout Yosemite.

LONG THE CENTER of activity within Yosemite, by 1918 the Old Village had become tattered and outdated. Growing crowds often clogged the single street through the village. Visitors grumpily stood in line waiting to get into the small headquarters building to register for a campsite. Making matters worse, the long lines of visitors looking to pay for a site spilled out into the street, blocking the village's main thoroughfare. When visitors did secure their permits, they often found campsites overrun with cars parked haphazardly and tents set up wherever campers could find an open spot. Trash and sewage littered the park, as privies overfilled and visitors tossed their garbage anywhere they could. The park's lodges fared little better. Long Yosemite's primary hotel, the Sentinel was showing its age and needed renovation. With summer crowds growing, the Park Service finally decided that the village's shabby condition could no longer be overlooked.[2]

By 1922 the Park Service decided to replace the ramshackle collection of old shops and lodges along the south bank of the Merced River with a new village on the river's northern shore. Park Service Director Stephen Mather assigned Harvard-trained architect Charles P. Punchard Jr. to design the new buildings, starting by replacing the old Sentinel Bridge and building a new schoolhouse for the children of future employees. Funding delays slowed further construction, however, frustrating Mather to the extent that he financed the construction of the Ranger Club himself. Located across from the new administrative building, the two-story wooden structure provided much-needed housing for Yosemite's bachelor rangers. The Club reflected Punchard's efforts to harmoniously blend each new building with the surrounding landscape by using local materials as much as possible, creating a rustic aesthetic for the village.[3]

Sadly, Punchard died after a long struggle with tuberculosis soon after designing the Ranger Club. His fellow architects Daniel Hull and Myron Hunt took on the task of completing the new village soon afterward. They began by designing the park's administration building and post office. Reflecting the popular Arts and Crafts architectural style of the time, each building embraced Punchard's vision by blending local materials and simple rustic design in creating what would later become known as "parkitecture." Nearly every national park used the look afterward, creating a distinctive feel for national park visitor centers, administration buildings, and comfort stations until the mid-1950s, when the Park Service embraced modernist architecture in its new buildings. To this day, however, the rustic architectural style remains a hallmark of national parks for millions of visitors.[4]

The new village became significant not only for its architecture but also for how it transformed how visitors experienced Yosemite; it was often their first stop in the park. After parking in the newly constructed lot with ample spots, they entered the administrative building, where Ansel Hall, a botanist with a forestry degree from the University of California and an information ranger since 1919, greeted them. After assigning them a campsite, and updating them on the weather, Hall suggested visitors tour the exhibits in the adjacent room to learn about the park before they headed out into it.

The small exhibit room quickly proved inadequate to hold the growing number of exhibits and visitors. Hall, who had been hired to enhance Yosemite's interpretive efforts and had written two guidebooks on the park, soon began advocating for the construction of a stand-alone museum where visitors could learn more about Yosemite's natural history, geology, and Native heritage. Funding for such a building proved elusive, however. While Congress was supportive of the newly established National Park Service, it remained rather stingy when it came to funding individual parks. To address the lack of federal funding, Hall organized the Yosemite Museum Association to raise the necessary money. After three years of effort, the association secured a sizable donation from the Laura Spellman Rockefeller Foundation, and the museum finally opened two years later. The squat two-story building included three galleries and a modern fireproof archive that safely housed Chris Jorgensen's Native basket collection, artwork by Thomas Ayres, and other priceless artifacts.[5]

Some 33,000 visitors passed through the museum's doors its first summer, in 1926. On entry they passed by exhibits on Yosemite's geologic history before walking into a small gallery on the park's natural history. But it was in the larger Indian Room that most spent their time. After pausing in front of a colorful diorama presenting a scene of Ahwahnechee returning home from a successful hunt, visitors moved on to admire a display of intricate baskets woven by Paiute and Miwok women. They chatted over the display of Native dresses before making their way to the central exhibit in the room, which told of the importance of the acorn to the Native way of life.

Leaving the museum, visitors entered a courtyard, where, in the shade of an enormous oak, Paiute artist Maggie Howard sat weaving a basket. As the crowd pressed in to watch her masterfully craft a basket from white willow, fern root, and deer grass, a ranger naturalist launched into a brief talk on the history of the valley's Native peoples. A Mono Lake Paiute, Howard had spent most of her life in the Yosemite Valley. The Park Service hired her to demonstrate Native crafts at the museum for $2 a day plus whatever revenues she made selling her baskets. Generations of Miwok and Paiute women, including Lucy Telles, Julia Parker, Maggie Howard, and Emily Dayhoff, continued this practice into the twenty-first century.

As the ranger concluded his talk, Native performer Chris Brown appeared costumed as Chief Lemee to dance and tell stories to the crowd. Brown was born in the Yosemite Valley in 1896 to a Southern Miwok mother and white father. His performances of dances and elaborate costumes, both of his own invention, became one of the most popular attractions in Yosemite, with visitors waiting hours to see him perform.[6]

Visitors came away from meeting Brown and Howard believing they had witnessed a part of Yosemite's authentic past. Some even got a photograph, though Howard often became upset if visitors did not ask before snapping a photograph or failed to pay her for the picture. But authenticity is a tricky thing. Brown, for example, purchased a beaded vest and moccasins and made a feather headdress after visitors complained that he did not look like the Indians they saw in Western matinees. That Sierra Natives never wore such headdresses did not matter. Brown's appearance as a stereotypical past tense "Indian" was all visitors wanted, or needed, to enjoy his performance. By taking a photograph with Howard or playing a game with Chief Lemee, visitors experienced what they truly believed to be a genuine connection with Yosemite's frontier past. Nowhere was this more apparent than Indian Field Days.

THE BLEACHERS SAT in the shade of black oaks on the edge of the meadow, a pleasant place to sit and enjoy such a warm August day. The dwindling Yosemite Falls sat as background as a troop of Boy Scouts attempted to ride, with little success, several donkeys. The stout beasts bucked hard, knocking off each boy in seconds. The scene was all rather humorous until a donkey took exception to the roughhousing and kicked a young scout. Rangers raced to the screaming boy, quickly bandaged his leg, and called an end to the event.

After seeing the injured scout off, Ranger Ansel Hall and his friend Carl Benson performed a skit as bungling campers. The sketch struck close to home for many in the audience, who howled in laughter as the two men struggled to start a campfire, and then got into an argument over how to best get the fire going before finally putting it out. Audiences cheered as Wild West Show performer Betha Blancett won a friendly horse race against Secretary of the Treasury William Gibbs McAdoo. That evening "Skeeter" Bill Robbins showed off his roping tricks. A rodeo star, the Wyoming cowboy and his wife Dorothy had just moved to Los Angeles, where he had found work as a stuntman in the movies while managing film star Hoot Gibson's ranch.[7]

Created in 1916 by Yosemite's chief ranger, Forest Townsley, Indian Field Days began as a way to attract visitors to the park during the slower fall months. Having grown up in Oklahoma, Townsley sought to recreate the powwows of his youth by focusing on Miwok art and culture. But the simple affair attracted a limited number of visitors in its first two years. To resolve this issue, the ranger expanded the event to include rodeo competitions, games, and other contests. Part rodeo, pageant, and craft fair, Indian Field Day quickly gained popularity, drawing thousands from across California every fall.[8]

Visitors paid handsomely for Native basketry and beadwork. Miwok women won cash prizes as large as $550 in competitions for the best basket. Natives from around the state competed in costume contests for the best cowboy and cowgirl, the chubbiest baby, and various field day events. Tourists likewise vied for prizes in tug-of-wars and Roman chariot race contests with park employees. Competitions like the Indian Baby Show both commodified and acculturated the valley's Native inhabitants. Yet, by the summer of 1922, Indian Field Days had grown so popular that the Park Service was forced to move the event to the broad meadow south of the Old Village in order to accommodate all of the visitors and performers.[9]

More than 24,000 attended Indian Field Days in the summer of 1929. The Park Service had promoted the event far and wide, even securing sponsorships from Standard Oil and the Yosemite Valley Railroad in attempts to turn a profit. Even with so many attending, the event lost an astounding $1,387.13, as proceeds fell far short of the costs of holding the increasingly popular event. Faced with such a deficit, along with the damage done to Leidig's Meadow, where the horse racing had taken place, Yosemite's advisory board cancelled Indian Field Days for good. The park saw a dramatic 25 percent drop in visitation the following year, which Superintendent Thompson directly attributed to Indian Field Days' absence.[10]

The Park Service's cancellation of Indian Field Days further strained the already tense relationship between the agency and the park's Miwok and Paiute residents. The event had served as an important source of income for many Natives living in the park, and its end meant a sizable hit to many families' budgets. Conversely, the Park Service increasingly saw the valley's Native residents as a burden. With no federal recognition, the Miwok had no legal right to remain within Yosemite. Only the Park Service's goodwill had allowed the small community to live in a collection of cabins and traditional bark houses, among granite boulders and ponderosa pines, at the mouth of the Indian Canyon. The Park Service understood the village's state distracted from the valley's beauty. Moreover, the agency falsely believed the village's Native inhabitants to be lazy and prone to drinking rather than working, even though most of the valley's Native population worked within the park, some for the past twenty years. Moreover, many were military veterans. But despite such truths, Yosemite's Native population remained on the fringes of the park, often visible to visitors only as part of Yosemite's historical past, not its present. Because of such views, the Park Service sought some way to remove the village's residents.[11]

Whooping cough tore through the Indian Village in 1930, giving Superintendent Thomson the excuse he needed to move the park's Native residents to a new home. Within a year Park Service landscape architects identified a location for a New Indian Village farther down the valley. Shielded by a grove of trees from visitors' view, the new village included cabins with cookstoves and toilets. Even with such modern conveniences, the Park Service failed to provide any parking for the residents, meaning they had to park their cars elsewhere in the valley and walk to their new home. Making matters worse, limits on who could live in the New Indian Village remained a point of contention between the Park Service and the valley's Native inhabitants.

The Park Service sought to reduce the number of Indians living within Yosemite by allowing only those who had descended from the valley's original inhabitants to live in the park. Only after it was pointed out that the decision would effectively remove many of those essential to the park's Indian cultural program, like Maggie Howard, did the Park Service relent and allow residents to stay in their homes. The Park Service finally razed the Old Indian Village in 1933, leading Thomson to celebrate having the "deserving Indian residents of the valley comfortably housed." However, the issue of Natives living in Yosemite remained unresolved and would lead to future conflicts over who deserved to call Yosemite home.[12]

DON TRESIDDER FIRST visited Yosemite as a twenty-year-old tourist in 1916. Over the next thirty years, he would become one of the most important figures in the park's history. Tresidder grew up in the rural Midwestern town of Tipton, Indiana. After contracting tuberculosis at age twenty, he followed his doctor's advice to spend a summer in Southern California's drier climes. His sister Oliene joined him for the trip across the country. Yet the siblings found themselves stranded in Fresno after heavy rains knocked out the tracks to Los Angeles. During their time on the train to California, they had befriended a fellow passenger who regaled them with tales of Yosemite's beauty. Delayed until the railroad was repaired, Don and Oliene decided to see Yosemite for themselves and boarded the next stage to the park. They were immediately enthralled by the park, on arriving, and spent the next several days touring the Yosemite Valley. When it became time to return to Fresno, the siblings chose instead to remain in the park for the summer, finding work with the Curry Company—a decision that changed both their and Yosemite's futures.

Working as a porter that summer, Don befriended Stanford mathematics professor Rufus Lot Green. Immediately impressed by Tresidder's poise and wit, Green suggested that Don attend Stanford rather than return to Chicago that fall for school. He even went as far as offering Tresidder room and board in exchange for working as a cook, a questionable trade as Tresidder was no cook. Nonetheless, the young student agreed and began his studies at Stanford that fall.

That same summer, Tresidder met and began dating the Currys' youngest daughter, Mary. The two quickly became inseparable, though Mary's father did try to intervene by temporarily firing Tresidder after the young couple had snuck off to climb Half Dome together. While often described as shy and bookish, Mary was an accomplished mountaineer who had spent much of her childhood exploring Yosemite's peaks. Like her parents, she had attended Stanford University. After earning her master's degree in English in 1916, she became one of the first women to attend Yale University. There, she earned a graduate degree in English poetry, was inducted into the Phi Beta Kappa honorary society, and began work on her doctorate. Mary never completed her doctorate, however, returning instead to Yosemite and to Tresidder. In June of 1920, the couple married in a small ceremony and spent part of their honeymoon exploring Yosemite's backcountry.[13]

Following a brief stint in the military, Tresidder completed his bachelor's degree at Stanford in 1919 and immediately enrolled in medical school. The couple's life came to center around Don's medical studies during the winter, and Yosemite during the summer. Tall, handsome, and driven, Tresidder rose rapidly within the Curry Company's management, soon becoming co-manager with his brother-in-law Robert Williams. Similarly, Oliene became manager of the Wawona Hotel before meeting and marrying San Francisco broker Lucio Mintzer in 1933.

Conflict over the much-sought-after ownership of a gas station within the Yosemite Valley, between the Curry Company and its rival Yosemite National Park Company, led Park Service Director Horace Albright to order the two concessioners to merge into a single company in 1925. Despite Tresidder's relative youth and inexperience, he was all of thirty at the time, Albright named him the Yosemite Park & Curry Company's (YP&CC) president. Given a monopoly within the park, the newly formed company oversaw the development of the Ahwahnee Hotel and Badger Pass Ski Area. Despite his new responsibilities, Tresidder finally earned his medical degree, three years after taking control of Yosemite's concessions, by completing his thesis on public health in the national parks. That same year, the YP&CC opened the opulent Ahwahnee Hotel.

Designed by the architect of Yellowstone's Old Faithful Lodge and future designer of the Grand Canyon Lodge and the Lodge at Sun Valley, Gilbert Stanley Underwood, the Ahwahnee sat just under the Royal Arches. It featured ninety-seven rooms and suites, a grand lounge, and an enormous dining room. Underwood borrowed from the park's rustic architectural aesthetic by using local stone and timber, blending it with art deco elements to create one of the most significant lodges in the National Park System. Visitors spent time on the hotel's patio, soaking in the view of Half Dome, before enjoying a black-tie dinner in the grand ballroom, then moving to the lounge before retiring for the evening. After the repeal of Prohibition in 1933, the YP&CC converted a private dining room into the El Dorado Diggings Bar, themed after the California gold rush.[14]

California's rich and famous flocked to the Ahwahnee. Will Rogers, Walt Disney, Shirley Temple, and Humphrey Bogart were but a few of the notable guests that stayed at the hotel. Additionally, President John F. Kennedy, Queen Elizabeth, and Eleanor Roosevelt each stayed at the Ahwahnee during their visits to Yosemite. But despite attracting the powerful and famous, the hotel struggled to attract guests during the winter. Few found the thought of visiting Yosemite during the colder, snowier months appealing, even after the completion of the All-Year Highway. To attract winter guests, Tresidder began holding a holiday celebration: the Bracebridge dinners. Based upon the Washington Irving novel *Bracebridge Hall*, the holiday feast became an annual tradition. In 1929 Tresidder asked family friend Ansel Adams to take over the dinner, to which the photographer readily agreed, remaining its director until 1973.[15]

AS DEVELOPMENT WITHIN Yosemite grew, the park came to experience many of the same growing pains as the nation's cities, particularly regarding pollution. Sewage quickly rose as one of the more pressing issues brought about by increased visitation. Yosemite's earliest hotels had addressed the problem by dumping their sewage into the Merced River, leading each new hotel to move farther upstream to avoid the stench of polluted water. Similarly, with no viable alternatives, trash was typically burned or buried. These methods remained in place until the U.S. Army took over management of the valley in 1906. Sanitation, noted then–Acting Superintendent Gabriel Sovulewski, took precedence over all other park issues. To manage human waste, Sovulewski restricted camping in the valley to meadows east of the Old Village. Although regular garbage collection was not established until six years later, the cavalry did infrequently use wagons to transport trash to a dump for burial. However, these solutions proved inadequate with the surge of automobile travel into Yosemite, compelling the Park Service to seek new solutions for its growing waste problem.[16]

Sewage remained a challenging issue, however. In 1920 the Park Service built a sewage plant near the Three Brothers rock formation, connected to both the hotels and campground via a series of ditches. While the plant greatly improved public health within the valley, visitors often complained of the stench it created during warm summer days. Faced with growing demands that it address the odor, the Park Service built a new plant farther from public view and smell north of Bridalveil Fall.[17]

But the treatment plants solved just one side of the growing problem. Beginning in 1923 the Park Service began building comfort stations throughout the park, particularly in heavily used sites such as campgrounds, entrance gates, and overlooks. The new bathrooms were first connected to the Three Brothers treatment plant, and then the newer facility, creating a hidden sewer system that linked the valley together in a web of sewage pipes, pump stations, and rain gutters. Appreciative of the modern toilets, few visitors put much thought into where their waste went. And the Park Service hoped to keep it that way.[18]

Even with the construction of a modern sewage system, garbage remained a problem, especially when it came to the park's bears. The crux of the problem lay in the fact that the park, bowing to pressure from its concessioner, continued to allow bears to feed on food waste in a large garbage pit. For a fee, visitors were allowed to watch the bears every summer evening from stands surrounding the pit under the watchful eyes of armed rangers. The spectacle acclimated bears and humans to each other, leading to increasingly dangerous interactions between visitors and bears.

Seeking an easy meal, bears often wandered into campsites. Thrilled to have such an encounter with a wild animal, visitors would feed the bears by hand, pat them on the head, or even load their children onto a bear's back. Not surprisingly, bear attacks grew, forcing rangers to intercede. Many visitors resented rangers' attempts to prevent them from feeding the bears, wrote Superintendent Lewis on the growing problem, and simply refused to leave the animals alone.

The solution to Yosemite's bear problem remained elusive for years. Lewis suggested dousing the pile of garbage in crankcase oil and setting it on fire. But visitors' complaints about the thick black smoke and stench made burning the trash not an option. Park naturalists pushed for a solution that did not resort to killing the bears. So, in the fall of 1929, park rangers began baiting hungry bears into wheeled traps and then moving the bears farther into the interior of the park. Those bears who returned to the valley were marked with paint, with repeat offenders eventually euthanized. Three years later, the Park Service finally shut down the bear feeding shows. While removing the popular attraction, the closure did reduce the number of human-bear conflicts within the Yosemite Valley. Yet the issue of how to best manage bears and visitors remains a struggle today.[19]

Bears were not the only animal attraction within the park. In 1918 rangers saved three orphaned mountain lion cubs. Rather than destroy them, park rangers decided to put the three on display near the new village. Other orphaned animals soon joined the trio, creating a small zoo in the valley. With the zoo's apparent success in entertaining visitors, the California Academy of Sciences convinced Yosemite to house a herd of tule elk, whose numbers were in decline in their native San Joaquin Valley. Despite protests from University of California zoologist Joseph Grinnell, Director Mather loved the idea, believing it would further bolster visitor numbers. And so, in 1921, twelve elk were released into a twenty-eight-acre paddock in Cook's Meadow. There, the elk thrived, quickly outgrowing the enclosure. Despite such pressures, the zoo and elk herd remained in the valley until 1932. The Park Service finally relocated the elk east to the Owens Valley and found homes for the remainder of the Yosemite Zoo's animals, before closing the facility permanently.[20]

THE FLOOD GATES closed on the O'Shaughnessy Dam in the summer of 1923. Within a month, the Hetch Hetchy Valley vanished beneath the waters of the new reservoir. Named after San Francisco's city engineer, Michael O'Shaughnessy, who oversaw the dam's construction, the towering plug of concrete had taken four years to build. Laborers had clear cut the valley of all its black oak and scrubs and carved a 20-foot-diameter tunnel to divert the Tuolumne River before constructing the 430-foot-tall dam. Standing on the dam, city and Park Service officials, San Francisco's former mayor James Phelan, who had just lost his reelection bid as the state's U.S. senator, and other dignitaries dedicated the dam's completion. After several speeches, O'Shaughnessy took the dais for the day's final words. "The O'Shaughnessy Dam and Hetch Hetchy reservoir stand as a refutation to so-called 'Nature lovers,' who opposed its construction," he opened. The reservoir located in a national park was the very embodiment, he argued, of where beauty and utility wedded. Those in attendance cheered, and San Francisco's dam was finally complete.[21]

Yet, the inundation of Hetch Hetchy did not end the controversy over the dam, or the fight over who ultimately controlled the Tuolumne watershed. From the moment the reservoir filled, the Park Service understood that San Francisco had no intention of acknowledging the federal agency's jurisdiction over the watershed, especially its obligations under the Raker Act to provide public access. O'Shaughnessy publicly refused to build the trails and recreational facilities required under the act. Rather, he sought to exclude public access to the Tuolumne watershed entirely in an effort to seize control of the watershed for the city. Doing so, O'Shaughnessy understood, would ensure the purity of its water source, as well as provide the basis of the city's hydroelectric power. All other uses were simply a waste of public tax dollars.[22]

Conversely, the Park Service insisted San Francisco's city government open Hetch Hetchy to visitors, starting with replacing the Hetch Hetchy Railroad with a roadway suited for automobiles. With the dam completed, the Park Service argued, the narrow-gauge railroad had simply outlived its usefulness, and repurposing the route for automobiles would better serve visitors by opening the northern half of Yosemite to further visitor exploration.

The Park Service did not stop there. They pressed the city to finish constructing a road around the reservoir's northern shore that would provide access to an expanded trail system and facilitate visitors' greater access to

Yosemite's vast backcountry. But the vision extended beyond mere infrastructure. The agency also envisioned three new campgrounds, a restaurant, and even a modern hotel, akin to those nestled in Yosemite Valley, along the reservoir's shores. Their ambition was not just to pave roads and build trails but to create a network of facilities that would transform Hetch Hetchy into a gateway for experiencing Yosemite's more remote corners.[23]

City officials rejected the Park Service's promised new campgrounds, restaurants, and lodges, citing the Raker Act's prohibition of bathing, washing, or watering stock within the reservoir's boundary. The city took a rather liberal view of the act's language, posting signs barring visitors from camping alongside or even swimming in the Tuolumne River above the reservoir. While the Park Service removed such postings as quickly as the city put them up, the struggle over who controlled the watershed remained heated. City officials viewed the watershed as their domain and stated that in order to guarantee its citizens pure drinking water and cheap hydroelectric power the entire region needed to be closed to public use. Not willing to cede a third of the park, Park Service Director Stephen Mather pushed back, tersely suggesting that if San Francisco was so concerned about its water quality it should do what other cities had done and simply chlorinate the water. City officials spurned the suggestion, incredulously wondering if the federal government placed the enjoyment of a few thousand over the safety of San Francisco's entire populace.

While largely siding with Mather on the topic of access, the public remained deeply skeptical of his assurances that he would not collude with San Francisco in excluding them from Hetch Hetchy. When news reports surfaced about an agreement between the city and the park barring all recreational use of the Tuolumne River, a deluge of letters protesting the decision flooded the superintendent's office. Visitors soon began encountering signs prohibiting camping and fishing along the river, seemingly validating the rumors. To many, these signs seemed like incontrovertible evidence of the park and the city's collusion to restrict access.

The alleged alliance between the Park Service and San Francisco in shutting off nearly a third of the park was not based in fact, but the perception was powerful. It reflected the tangled and often perplexing relationship between federal authorities and city officials over control of the watershed, as well as the public's conviction that Yosemite was for all to enjoy. The situation reached a boiling point when rumors began to circulate about city officials fishing behind locked gates.[24]

Six cars pulled up to the gate at the O'Shaughnessy Dam entrance early on Sunday, May 29, 1927. A passenger jumped out of the lead car, jogged to the locked gate, opened it, and stood aside as all six vehicles pulled through. As the last car drove through the open gate, he closed the gate, locked it, and jumped back into the lead car. The convoy then slowly headed up the narrow dirt track to Lake Eleanor. It was not the first time city employees had used their position to gain access to the remote lake. Standing below, a member of the Pacific Rod and Reel Club angrily watched the whole thing in disbelief.

Located north of Hetch Hetchy, Lake Eleanor was an angler's dream. Its stunning clear waters were full of trout unused to the fisherman's cast. For those city employees who could claim to have business at the lake, even if it was known that that "business" was catching a few trout, the lure of a private fishing paradise proved too alluring to miss. But to those left outside the gate, the city employees' brazen act confirmed beliefs that the city was violating the law in keeping them locked out of one of the state's prime fishing spots.[25]

In a tersely worded letter to Yosemite's superintendent, the Pacific Rod and Reel Club of California's chairman, Gilbert Jones, protested the city's restrictions on camping and boating on Hetch Hetchy and access to the road to Lake Eleanor. "In as much as this is within the National Park, we assume that Jurisdiction [*sic*] over all matters in connection therewith would rest with Park Officials, and not with local State Officials," wrote Jones. Believing that the city held no authority to restrict access under the Raker Act, Jones offered to violate the restrictions in order to decide the issue in court. "Our organization feels that the use made of this Park Territory has been violative of the whole principle upon which the National Park is established," he concluded, and so must be stopped.[26]

Jones's letter ultimately made its way to Director Mather, who, while sympathetic to Jones's argument that the city held no jurisdiction over visitors' recreational use of Hetch Hetchy, decided after discussing the matter with the secretary of the interior that it was an inopportune time to press the city on the topic. Rather, he would bring it up during his visit with city officials later that summer. Furthermore, Mather ordered that Lake Eleanor Road remain closed to all users until it was made safe for automobile travel and stationed a ranger to keep all anglers, whether they worked for the city or not, from accessing the lake.[27]

Despite his best efforts, Mather admittedly did not get very far with the San Francisco Board of Supervisors in getting the city to pay for the promised trails and campgrounds and opening Lake Eleanor to the public. Instead, the Board took the opportunity to play to its constituents on how they were trying to save their tax dollars. Furthermore, the Board opposed the very idea that control over the land surrounding both reservoirs should be returned to the federal government after the city had paid for the improvements.

Costs also played a central role in the battle over who ultimately controlled Hetch Hetchy. Already upset about the federal government's lack of economic support in the reservoir's construction, San Francisco's elected officials were increasingly outraged at the Park Service's insistence that the city pay for the construction of new roads and trails. On hearing Mather's demand that the city pay $1,628,500 to meet its obligations in building a road to the dam, along with the widening and paving of the road to Lake Eleanor, an assistant city engineer exploded that there was no need for such improvements when a simple wagon road would suffice in servicing the dam.[28]

The fight over Hetch Hetchy was not only over water quality, or even taxpayer costs, but more importantly over who controlled the future of the national parks. If San Francisco was successful in carving the Tuolumne watershed out of Yosemite, then the future of the National Park System, and the national park ideal, would be put into doubt. But if Yosemite remained intact with the Park Service in control, then recreational use of public lands remained as valuable as extractive wealth. It was a fight the nascent federal agency could not afford to lose. And so the political knife fight over the future of a remote watershed high in the Sierra Nevada became an issue of national importance. So, in October of 1927 three U.S. congressmen traveled to Yosemite to voice their support for the park in its ongoing struggle with the City of San Francisco.

"The failure of San Francisco to perform its obligations under the Hetch Hetchy Act, and the frequent attempts of its officials to interfere with the enjoyment and use of the park by the public, is about to become a national scandal," Michigan Congressman Louis Crampton roared during a small luncheon at the Ahwahnee Hotel. Speaking for his fellow representatives Burton French of Idaho and Colorado's Edward Taylor, Crampton launched broadside after broadside at San Francisco's efforts to close Hetch Hetchy to the public, attacking the city's failure to live up to its side of the bargain that had allowed the construction of the dam within Yosemite. After the congressman's speech, Director Mather stood up and noted that the tension over Hetch Hetchy lay with the lack of cooperation from San Francisco government officials, and not with the city's residents, many of whom had contributed to Yosemite's establishment and had enjoyed the park.[29]

Even with Congress's involvement, the struggle over the Tuolumne watershed continued for another decade. The city and Park Service briefly toyed with building a road connecting the Tioga Road to Hetch Hetchy. But after negotiations fell through, the idea was scrapped and the Park Service focused on developing the road between Camp Mather and the reservoir. Finally, in 1938, Park Service and San Francisco officials agreed the city would pay the park the costs of maintaining the Hetch Hetchy Road, as well as the needed sewage treatment in Tuolumne Meadow. In exchange, the Park Service would restrict visitors' access to the reservoir. Bathing and swimming remained prohibited, as was boating, though fishing from the shore was allowed. The Park Service concluded that improving the road to Lake Eleanor was far too challenging, so they closed it to all automobile travel and converted the route into a hiking trail. Sections of the old roadbed remain apparent, but nature has reclaimed much of the former route.[30]

Hetch Hetchy remains a point of contention today. Environmental advocates contend that the removal of the O'Shaughnessy Dam will restore the Hetch Hetchy Valley to its past splendor. Yet their efforts are unlikely to succeed, as the City of San Francisco remains resistant to giving up the crown jewel of its waterworks, particularly as climate change threatens future water supplies. Few visit Hetch Hetchy as it is distant from the rest of Yosemite. A Fish and Wildlife Service survey in 1975 counted a mere 1,101 visitors between April and November of that year. A National Park Service study seven years later estimated that if the valley were returned to its pristine beauty, upward of 50,000 would visit Hetch Hetchy annually. That number appears rather conservative when compared to the millions that visit the park every summer, and the draw a reclaimed Hetch Hetchy would have to visitors. Needless to say, such discussions remain speculative, and the loss of the Hetch Hetchy Valley remains an open wound in Yosemite's history.[31]

MARY AND DON Tresidder returned from Switzerland in 1928 with a new vision for Yosemite. The couple had attended the inaugural Winter Olympic Games in St. Moritz during their European tour. Inspired by the jubilant crowds and the quaint Swiss village's stunning scenery, the owners of Yosemite's now only concessioner imagined transforming the park into a similar winter wonderland. Such a dream was not as outlandish as it may have appeared. The All-Year Highway's completion was already drawing tourists to the Yosemite Valley during the winter in numbers unimaginable just a few years prior. Additionally, the Ahwahnee Hotel offered first-class amenities like those the Tresidders had enjoyed in St. Moritz. But it would take more than a luxury hotel and a highway to host the world. Long known for its mild climate, California would need to change its image and develop a vibrant winter sports scene, and Yosemite would need facilities worthy of Olympic competition.

Within months of returning, Don launched the Yosemite Winter Club. Aimed at cultivating winter sports in the park, the Club quickly gathered a distinguished membership, including many of Tresidder's notable friends from Los Angeles and San Francisco, as well as Park Service officials and most of Yosemite's year-round residents. But the Winter Club was much more than a social group. It sought to build a broad embrace of winter recreation throughout California, and in doing so signal to the International Olympic Committee that Yosemite was suitable as an Olympic venue. Moreover, winter use of the park would also attract visitors year-round, no doubt bolstering the YP&CC's bottom line and continued construction of the new village.

The company converted the Camp Curry parking lot into a large ice-skating rink, drawing crowds eager to skate beneath the majestic backdrop of Glacier Point. Enthusiastic skaters engaged in hockey matches, speed skating races, and curling competitions. Stoneman Meadow became a hub of winter activity, offering dog and horse sleigh rides alongside snowshoe and skijoring races. The YP&CC even constructed a modest ski jump on the moraine near the horse stables. Yet the most thrilling attraction was the toboggan slide just west of Camp Curry. From its lofty perch, visitors clambered into a sled and raced down the banked course at exhilarating speeds, only to come to a thrilling halt at the bottom. Of course, the thrills sometimes turned dangerous, as was the case when Berta Beers's toboggan slammed into another guest's sled after they failed to get it off the track in time. Spilling off her sled, the young woman landed headfirst into a railing. While she suffered only minor injuries, Beers's accident was just one of several violent crashes on the popular ride, helping lead to its eventual closure in the 1950s.[32]

The Park Service shared the Tresidders' vision of bringing the Olympics to Yosemite, lobbying both state and Olympic officials on the national park's appropriateness as an Olympic venue. Winter tourism remained a goal for both the federal agency and its concessionaire, and the Olympics promised to deliver just that. Yet after a year of frantic lobbying, the International Olympic Committee awarded the 1932 Winter Games to New York's Lake Placid. Dismayed but not disheartened by the decision, the Tresidders continued to build Yosemite into a winter destination.[33]

The Olympics were not the only idea the Tresidders brought back with them from Switzerland. While in the Alps, the couple had fallen in love with downhill skiing. They were not alone. By the start of the decade, thousands of Americans across the country were taking up the sport. The YP&CC responded by building a small ski hill next to its stables. While this was popular, visitors complained about the hill's gentle slope, inconsistent snow, and location. One particularly peeved skier noted that the hill ended on the road, making it particularly challenging for beginners to stop before skiing into traffic—and that was if they made it past the stand of trees at the trail's bottom, which many failed to avoid. The Park Service agreed with such complaints, noting that the ski hill was "too short and dangerous" for novice skiers.[34]

With a limited number of suitable hills for skiing within the valley, the YP&CC embarked on a search for an ideal site to establish a ski operation elsewhere in the park. The primary issue was that snow rarely stuck to the ground below seven thousand feet, within the Sierra Nevada, for more than a handful of days at a time. This meant that any new ski hill would need to be located farther away from the Ahwahnee. Glacier Point was the most likely spot, except few wanted to make the hike up the Four Mile Trail during the winter. Moreover, the Glacier Point Hotel closed during the winter. The idea of a tramway garnered initial interest from both the concessioner and the Park Service, but it was eventually abandoned due to its high costs.

The completion of the Wawona Tunnel in 1932 resolved the issue. Carved through several tons of granite, the new tunnel not only created one of the park's most iconic overlooks but also reopened the old Wawona Turnpike route to winter drivers. Skiers were quick to discover the snow-covered slopes near the Chinquapin gas station. Situated midway between Wawona and the Yosemite Valley, the former settlement soon became an impromptu destination for an increasing number of California skiers.

However, the surge of cars quickly led to parking problems. Within the first year, skiers were forced to park alongside the road and trek half a mile up its icy shoulder. Once there, they crowded onto the few roughly cut trails, colliding with each other and the surrounding trees. The situation was exacerbated by the lack of privies, and only the single twelve-by-fourteen-foot construction shack provided skiers shelter from frigid winter temperatures. It was clear to both the Park Service and the YP&CC that a better solution was needed. How to find that solution, however, was a point of debate between the park and its concessionaire.

But while the superintendent's office and YP&CC debated how to best solve the need for a better ski hill, visitors discovered the slopes above Monroe Meadows. Located just a few miles east of the small settlement of Chinquapin, along the road to Glacier Point, the area's steeper north-facing slopes were ideal for skiing. Skiers could easily hike to the highest point and enjoy the panoramic views of the Sierra Nevada, before schussing down the conifer-lined slopes to an open meadow below. Tresidder took note of the crowds venturing to the area and, within months, had the YP&CC shuttling skiers to Badger Pass. Still, the problem of proper facilities remained. What few toilets existed in the area were primitive at best, and the small shed the Park Service had provided in the meadow was often overcrowded and smelled of wet wool.

To solve the problem, in the fall of 1935 Don Tresidder submitted plans to build a ski lodge, rope tow, parking lot, and several new trails at Badger Pass. While supporting the development of skiing in Yosemite, Superintendent Thomson rejected the plan, fearing it would forever lock the park into maintaining a ski area, an inclination that proved to be true. Thomson was also exasperated by the YP&CC's expectation that the Park Service would build a ski area from which the company could reap the profits. Tresidder was equally frustrated with Thomson, convinced that the Park Service had committed to building Badger Pass as part of the promised "intensive promotion of winter sports" during the bid to secure the Winter Olympics years earlier. Moreover, he adamantly disagreed with Thomson's argument that the concessioner's plan was far too hasty, and that further study was needed. In a letter to Thomson outlining his case for the ski area, Tresidder wrote that he knew of no sport attracting greater interest in California than skiing, and with the Park Service's partnership Badger Pass would soon become the premier ski destination throughout the West.[35]

Thomson ultimately relented, and construction on the new ski hill began. Designed by San Francisco architect E. T. Spencer, the small ski lodge provided a warm place for skiers to rest and included indoor plumbing. Civilian Conservation Corps (CCC) laborers cut and cleared several new trails and built an enormous counterbalance sled named the "Queen Mary," in honor of Mary Tresidder, that carried up to 160 skiers per hour to the top of the ski hill. Even with the increased capacity, lines still stretched to half an hour. A rope tow was later installed to alleviate the wait, yet skier numbers continued to rise, even throughout World War II. By the end of the decade, Badger Pass attracted more than 45,000 skiers every winter.[36]

Badger Pass was among a growing number of ski areas built within the national parks. Ten parks—Yosemite, Yellowstone, Glacier, Crater Lake, Lassen, Olympic, Rocky Mountain, Rainier, Sequoia, and Mount McKinley—built lift service skiing. Yet questions on the appropriateness of lift-serviced ski areas within the national parks slowly closed most of these areas. Only Badger Pass and Hurricane Ridge in Olympic National Park remain in operation.[37]

BADGER PASS'S SKI trails were just one of the CCC's lasting legacies within Yosemite. Over the course of seven years, enrollees labored throughout the park, replacing the cables on Half Dome, cutting road banks, treating blister rust, building campgrounds and trails, planting trees, and removing invasive plants. Created in 1933 and lasting until 1942, the federal jobs program sought to solve, in part, the nation's crippling unemployment by putting young men to work on public projects, from bridge building to firefighting. The tough physical work provided the needed labor to transform Yosemite into a modern recreational wonderland, laying the foundation for the postwar boom in visitation, as well as the environmental movement, through introducing the national parks to thousands of working class men, like Clair Nelsen, who later returned after the war to enjoy the fruits of their labor.[38]

Becoming a woodsman was almost too good to be true for Nelsen. He had just lost his part-time job as a clerk in a Bay Area grocery, and news of President Roosevelt's plan to put hundreds of thousands of young men to work in forest conservation thrilled the young Californian. He quickly applied and soon found himself reporting to Fort Barry, just outside San Francisco, for enrollment in the Civilian Conservation Corps. Vaccinated and clothed, he joined two hundred other men in forming one of the first CCC companies in the state. In May of 1933, the company loaded a train and headed into Sierra. After arriving later that afternoon, they loaded into trucks, and they were soon caught up in the spectacular scenery along the winding Sierra road. Finally, after a several-hour-long drive, the men arrived at a new camp in Buck Meadows, and they immediately began to finish construction of their new home.

Nelsen lived at the Buck Meadows Camp for the next two years. Like all CCC enrollees, he earned $30 a month, sending $25 of it back home. He ate well, gaining nearly fifteen pounds the first month at the camp. Sunday breakfasts quickly became his favorite meal of the week. They included eggs, bacon, ham, sausage, hot and cold cereals, pancakes, coffee, milk, and jam.

Seeking more adventure, and more pay, Nelsen volunteered as a firefighter. He soon gave up any romantic notions he had of the job, after spending days battling a particularly furious blaze. When there were no fires to suppress, Nelsen and the rest of his crew were put to work clearing brush. It was a hot and dusty job, but at least his life was not at risk. Within a year he became a truck driver and transferred to a crew building the Foresta Bridge just outside of Yosemite. He returned in 1981 to find the bridge still standing, if a little worse for wear.

But it was not just the work that forever changed Nelsen. With Yosemite only an hour away, many CCC enrollees spent their weekends exploring the park. On Friday and Saturday nights, Nelsen and friends loaded into one of the Forest Service trucks and drove to the valley. There, they danced the evenings away while wondering how they had become so lucky as to be in such a place. Nelsen was eventually hired by the Forest Service as a seasonal forester, a job he would have loved to have kept. But the world had other plans, and after serving in World War II Nelsen attended college and became a history professor at California State University Fresno, where his photograph still hangs in the halls of the Social Science Building. A scholarship in his name is awarded every year to a Fresno State history student. Following retirement, he continued to hike and explore Yosemite National Park, which had been introduced to him as a young man working for the CCC.[39]

A LIGHT WINTER rain on the morning of December 9, 1937, soon turned into a torrential downpour, and by afternoon it had swelled the Merced River. By the next morning, flood waters had washed out the Cascade Bridge and forced the CCC Cascade Camp's evacuation. Nearly a foot of rain fell over the next several days, washing out roads, destroying whole sections of trails, ripping buildings from their foundations, and wreaking havoc on the park's utilities. Happily, no lives were lost. However, the flooding caused an estimated $376,000 to $476,000 in damages and closed a large portion of the park for weeks afterward.

Yet the flood offered an opportunity to finally remove the Old Village park, superintendent Lawrence Merriam believed. With the new village having been completed two years prior, the Park Service wanted to demolish the remaining buildings within the Old Village. Advocates for the preservation of the Old Village buildings, particularly the Cedar Cottage, argued that the structures were the last physical evidence of the valley's pioneer past, so they deserved preservation. A lack of funding kept the park service from acting on its planned demolition, effectively preserving the village for another twenty years before the park finally razed the Old Village store and Degnan's Restaurant in 1959.[40]

It took three years to rebuild from the flood, during which time visitation dropped by nearly a third. But it was more than just the destruction wrought by the flood that kept visitors away. The nation's continued economic depression had slowed tourism to nearly a trickle nationwide. Yellowstone and the Grand Canyon saw similar drops in visitor numbers. Yet the Park Service remained optimistic that visitors would return to Yosemite, citing tourists attending the San Francisco World's Fair rather than coming to the park as the cause of the drop in numbers, not the nation's economic woes. Such optimism proved misguided as world events would soon throw the nation into chaos.[41]

Like the rest of the nation, Yosemite residents were stunned by news of Japan's attack on Pearl Harbor. Reflecting on the day, Yosemite's superintendent, Frank Kittredge, remembered that it became immediately clear that the coming war would forever change the park. Within months of the attack, many of the park's rangers, naturalists, and clerks had joined the military, while others had left to work in California's war industries. The loss of some of its most experienced men crippled the park's ability to operate at even a limited capacity, causing further tension between the Park Service and its concessioner. Faced with declining visitor numbers, the YP&CC's president, Donald Tresidder, laid off most of his staff, keeping only a skeleton crew on to run a handful of camps, lodges, and other facilities. Camp Curry, Yosemite Lodge, and Camp 16 remained open, while the Wawona Lodge and the Big Trees Lodge were shuttered. Glacier Point Mountain Home welcomed guests only during the summer, and then on a much-reduced basis. The Ahwahnee remained open to tourists until 1943, when it became a naval hospital. More problematic, Congress slashed the Park Service's budget, deferring needed maintenance and curtailing all operations within the park. Only three of the five campgrounds within the valley remained open, with one reserved solely for military personnel and their families. And while the Yosemite Valley Museum remained open on a limited schedule, the Mariposa Grove Museum was closed indefinitely.[42]

Within two years of the United States joining the war in Europe and the Pacific, rationing and travel restrictions caused visitor numbers to plummet from record highs to levels not seen in decades. Yet, even with far fewer visitors entering the park, congestion within the valley remained a pressing concern. Yosemite Advisory Board member John Buwalda noted that the thousands who did come to the park "scarcely get off the (valley) floor at all." With only three of the five campgrounds in the valley remaining open and as many as 11,000 campers descending upon the Yosemite Valley during summer weekends, visitors often found themselves standing in line waiting for camping spots rather than enjoying the park. Automobile trailers crammed in next to tents, leading to complaints of overcrowding. When visitors did venture out of their cramped campgrounds, they found many trails choked with downed trees and washed-out creek crossings due to a lack of trail crews to clear and maintain them. Yet visitors continued to come, even if it meant breaking a locked gate.[43]

A locked gate and snowdrifts had not deterred a family from entering the park over the Tioga Road in the winter of 1942. After breaking the locked gate on the eastern entrance to the park, the father of two somehow plowed his car through the early winter snow, making it as far as Crane Flat, where rangers waited for him and his family. After a rather terse conversation in which the man explained that he was simply trying to drive through the park, the rangers decided that the man would drive back to the broken gate, repair it, and then return to pick up his wife and two children at the ranger station before being allowed to continue westward. On the drive back to the gate, the visitor became so enthralled with the mountains that, upon his return to his family, he decided to spend the next several days in Yosemite.[44]

Not all misadventures included violating the law. Servicemen and women on leave continued to flock to Badger Pass to escape the war. Among them was naval officer Warren Miller. A lifetime surfer, Miller was admittedly out of his depth when it came to skiing. But the images of skiers schussing down snow-covered slopes enthralled him. Earning a well-deserved month-long leave after his subchaser sank in a typhoon off the coast of Guadalcanal, the young officer rented pairs of skis and boots and drove to Yosemite. His trip started inauspiciously when he nearly burned down his rental cabin attempting to wax his skis. Once he had made it to the slopes, Miller discovered that he had missed the wax so badly his skis refused to glide; instead, snow froze onto their bottoms, forcing him to scrape it off after every run. Making matters worse, his new ski boots did not fit, leaving his feet badly blistered. All in all, a poor start for one of the most prolific and beloved ski filmmakers of the next half century. Widely considered the godfather of ski movies, Miller created annual films that came to define the industry and introduced millions to the sport. But not all visits to Yosemite were filled with hijinks like Miller's. Many veterans and their families found peace from the war in Yosemite's silent wonder.[45]

The March afternoon promised to be mild as Frank Ewing began his hike from Happy Isles to Vernal Falls to inspect the trail's condition. The going was slow. Much of the trail remained covered in snow, and the winter runoff had washed out several sections. Believing that such conditions had kept most visitors from making the hike, Ewing was surprised to come upon an older man sitting alone at Clark Point. Having seen no sign of other footmarks in the snow, Ewing approached the lone hiker and asked how he had gotten up to the overlook. The elderly gentleman explained that he had made the hike up the Mist Trail with his wife and son, and that he was now waiting for the two to return from the top of Nevada Falls. Noting that the valley was relatively empty, as it was March and the cost of fuel was high due to war rationing, Ewing wondered how the family was enjoying their time in Yosemite. "This trip is not for me," the man answered, "it is for my son who is a sailor in the U.S. Navy and has a furlough for a few days." The clearly grieving father went on to tell Ewing that the couple had had another son who had died on the USS *Arizona* in Pearl Harbor, and that this may be the last time the three would have with one another. "I want to get him away from the war," the father sighed before Ewing said goodbye and headed back to the valley floor.[46]

On reading Ewing's report on the encounter, Yosemite's Superintendent Kittredge penned a brief memo to national park Director Newton Drury retelling the story. "If the trails of Yosemite National Park can help in the stabilizing of the mental condition of parents who have lost their sons in the war or can help to give a clearer perspective of the things worthwhile and the things worth fighting for," he wrote, "the parks are serving a grand purpose." Such sentiment underlined the National Park Service's search for its role during wartime.[47]

Citing Ewing's encounter with the grieving father as an example, Drury countered efforts by timber, grazing, and even mining interests to exploit national parklands during the war, maintaining the Park Service's central obligation was to preserve the national parks for future generations as well as provide solace for the war-weary nation. He wrote in the magazine *American Forests*, "In war, no less than in peace, the national park areas serve as havens of refuge for those fortunate enough to be able to visit them." He continued, "Affording an environment that gives relief from the tension of a warring world, the parks are, even now, being looked upon as a factor in the physical and mental rehabilitation that will be increasingly desirable as the war progresses."[48]

Yosemite came to play a much different role than a tourist destination beginning in June 1943, when the Navy commissioned the Ahwahnee and its surrounding grounds as a U.S. Naval Convalescent Hospital. Medical facilities, particularly along the West Coast, were quickly becoming overwhelmed with casualties from the Pacific, forcing the federal government to seek alternative facilities to rehabilitate servicemen recovering from the trauma of war. The navy paid the YP&CC a sum the company complained was too low, but the money allowed the company to cover expenses, such as taxes and upkeep, that with falling visitor levels they were unable to meet.[49]

While their beauty salved the burdens of a war-weary nation, the Yosemite Valley's towering granite walls quickly proved unsuitable for patients suffering from post-traumatic stress. The valley's towering walls caused sailors who had survived sinking ships to panic. Realizing their mistake, naval doctors quickly removed those suffering from such claustrophobia and reassigned the hospital to admit surgical and general medical casualties.

Despite its beauty, sailors came to complain loudly about the park's lack of diversions, save those readily available to tourists. And while many did enjoy renting a horse from the concessioner stables, playing a few rounds on the hotel's nine-hole golf course, or spending a day learning to ski at Badger Pass, such activities required money and were only novel the first several times. The single loudest complaint was the lack of alcohol and women within the park. For young men, many of whom were combat veterans far from home, such amenities were seen as indispensable.

Alcohol in particular was a constant source of tension between the Park Service and the seamen, particularly over the park's rules limiting access to when and where alcohol could be bought. When a ranger informed the men that the Ahwahnee Hotel bar was closed on Sunday, one sailor responded indignantly, "In Chicago you can buy it anytime." After receiving too many complaints from visitors over public drunkenness, Superintendent Kittredge shut down the bar entirely. Sailors responded by having visitors smuggle liquor into the park. Kittredge finally relented and allowed a small bar to open outside of the hotel. Sailors quickly named it Frank's Place, as a homage to the superintendent.

Like much of America, women volunteered to host social gatherings and dances for the seamen within the valley. One American Legion Dance attracted some two hundred men, many of them sailors, with one winning

$5 worth of portraits from Best's Studio in the evening's raffle. While most of these events allowed young men and women to simply mingle, a few led to romance. In July 1945 Camp Curry waitress Ann Marchetta from Jersey City, New Jersey, married Seaman First Class Frank Hammons of Buffalo, New York, in the naval hospital chapel. The couple had met while Frank had been on leave in Yosemite. Not all stories were as happy though. In December of 1944, a sailor was arrested and later convicted of rape committed within the park. Thankfully such violent crimes remained rare. The majority of the servicemen hospitalized in Yosemite enjoyed their time in the park so much that they returned with their families in the years following the war's end.[50]

As the persistent problems of overcrowding and access in the park did not altogether disappear during the war, neither did the animosity between the City of San Francisco and the Park Service over visitor access to Hetch Hetchy and Lake Eleanor. Mere days after the attack on Pearl Harbor, city officials reached out to the Park Service with concerns about the possibility of saboteurs damaging the O'Shaughnessy Dam and asked permission to arm its guards at the site. Superintendent Kittredge readily agreed, and within weeks the city equipped its eight employees with .30-caliber Winchester rifles.

However, questions over such a small force of untrained men being up to the task of protecting the city's water supply quickly led to the stationing of forty-three California state guardsmen at Camp Mather. Together with the park's few remaining rangers, the garrison closed the Hetch Hetchy Road in 1942, restricting all visitor access to the Hetch Hetchy and Lake Eleanor reservoirs to authorized personnel for the remainder of the war. The stationing of armed troops at Camp Mather effectively achieved San Francisco's long-held goal of closing the dam site to all visitors, though this time under the threat of a machine gun nest. While acknowledging the action was drastic, Assistant Superintendent John Wosky contended the troops offered the simplest means of protecting the Hetch Hetchy and Lake Eleanor Dams against fears of sabotage.[51]

But as concerns over security faded following the war, the struggle over recreational access to the watershed remained. San Francisco officials remained opposed to allowing any access in or around the reservoir beyond a small number of anglers fishing from the reservoir's rocky shore, believing that permitting visitors to swim or boat in Hetch Hetchy would irrevocably pollute the city's water supply. Visitor access, however, was a secondary point of conflict between the Park Service and the city. The Park Service's opposition to the city's proposal to sell excess electricity produced by the dam on the open market became front page news as the two governments continued to fight over control of the reservoir.[52]

AFTER MORE THAN a decade of financial struggles, the Yosemite Valley Railroad made its final run in August of 1945. Unlike the train that had brought Gladys Apker to the park twenty years before, the last run was a humble affair. A fire caused by the careless toss of a cigarette had destroyed the trestle at the small community of Bagby, far downriver from El Portal, earlier that summer. So rather than a line of Pullman cars full of tourists, or a flatbed piled high with logs, Train Number 23 made the much shorter run from Merced to Merced Falls to pick up two railcars. It was an inauspicious end for the once popular railroad. Having lost the majority of its passenger service after the opening of the All-Year Highway twenty years prior, the rail line had relied on freight to remain afloat. But transporting timber and other goods proved far too little to pay the bills, and in 1935 the Yosemite Valley Railroad filed for bankruptcy.

Disaster struck again when, just as the newly reorganized company began to turn a profit, a flood wiped out thirty miles of track, casting the railroad further into debt. The closure of the Yosemite Sugar Pine Company and sale of the Yosemite Portland Cement Company drove the final nail into the small railroad's coffin. Out of cash, and buried in debt, the railroad's mortgage was sold, and soon afterward the State of California granted the new owners approval to abandon the line.[53]

Nevertheless, the allure of rail travel to Yosemite continued to appeal to those seeking a solution to the park's traffic issues. The idea of returning to rail sparked the imaginations of many beginning in the 1970s. Writing in *National Parks and Conservation Magazine*, University of California, Santa Barbara graduate student Alfred Runte reasoned that, in the face of overwhelming numbers of cars entering the park, returning the railroad to Yosemite would guarantee maximum protection with maximum visitation.[54]

Runte was not alone in advocating a return to rail as the possible solution to Yosemite's overcrowding. In 1977 Sierra Club Books published designer and writer Christopher Swan and City of San Francisco professor Chet Roaman's novel *YV88: An Eco Fiction of Tomorrow*. Part science fiction novella, part illustrated schematic, the book proposed a radical transformation of how visitors enjoy Yosemite by removing all the roads into the park and replacing them with a solar-powered railroad that carries visitors from Merced into the Yosemite Valley. Written primarily by Swan, the book opens with a short tale of journalist Thomas Sutter riding the first solar-powered train into the valley in the summer of 1988. Along the way he talks to fellow travelers about how odd and wonderful it is to experience the park without the sounds of cars, crowds, or other distractions. It is an alluring story, at once fantastical and possible.[55]

Swan and Roaman posited Yosemite faced a three-part problem, reliance on the automobile, the erosion of the park's environment, and the concentration of visitors within the valley. The authors' solution, laid out in wonderful detail, was the removal of all roads and buildings from within the valley. It would be replaced by bucolic campsites served by a solar-powered train that would carry visitors from Merced to the Yosemite Valley floor, stopping at a lodge along the way. Having taken thirteen months to write, *YV88* failed to garner the immediate attention Swan and Roaman hoped for, and the book was never reprinted. Yet over the years it has gained a cult following, including many in the Park Service. And the idea of replacing automobiles with a train has gained attention, particularly in the face of climate change. Whether visitors would be willing to give up their cars and return to the days of rail remains an open question but does provide a possible solution to Yosemite's traffic problem.

AT THE WAR'S end, fears over the changing nature of the Yosemite Valley gained attention as the number of visitors increased. Complaints over the loss of the valley's meadows led the Park Service to study how the valley had changed since the invasion of white settlers in the 1850s. Published in 1943, the report concluded that the valley's meadows had decreased significantly over the previous century, and only its plowing, heavy grazing, and intentional clearing had prevented trees and shrubs from overrunning the valley floor. The report's authors blamed the removal of fire, which they noted Native peoples had used to manage the area for generations, combined with Galen Clark's dynamiting the Merced River moraine and heavy use by tourists, for the loss of scenic views. This loss contradicted the Park Service's mandate to preserve the natural beauty for future generations, the report's authors noted, and required much more active management in order to preserve the meadows for the benefit of those who had come a long way to see the wonders of the valley.[56]

Members of the Advisory Board echoed the call for the Park Service to take a more active stance in managing the park. In a series of meetings and field trips through the park, members William Colby, John Buwalda, and

Duncan McDuffie concluded that a number of issues needed to be addressed if Yosemite were to retain the feel it had prior to the war. Along with recommending halting the further encroachment of trees into the valley's meadow, the group warned that allowing crowded camps, congestion of cars on roads and pedestrians on paths and trails, and the generally noisy, bustling urban atmosphere would further diminish the public's enjoyment of the valley. Voicing their desire to focus on efforts to preserve the upper Yosemite Valley floor from increasing development, the Board members called for limiting the number of visitors allowed to overnight within the valley to current levels. Furthermore, the Park Service razed the Old Village and returned the ground on which it had stood to its native condition, as well as removing unnecessary commercial activities outside the valley. Such actions would allow current levels of use while preserving the valley's scenic beauty, they argued.

Of central concern to the Board were the growing crowds within the Yosemite Valley. Even with travel restrictions in place, many visitors came away disgusted with the park's congestion, unable to fully enjoy Yosemite unless the number of people in the park was considerably smaller. To lower numbers, Board members offered a half dozen suggestions. The first was to allow the overcrowding conditions to reduce the number of people who wanted to visit. If the park earned the reputation of being too crowded, the reasoning went, then people would simply quit coming. Or the Park Service could more aggressively seek to expand the travel season beyond the summer months.

While winter use continued to grow, this option faced the challenge that most visitors took their vacations during the summer months, when schools were out and the weather was much more appealing. The third suggestion was to eliminate entertainments like the Firefall, swimming pools, and golf courses. The elimination of such activities, unfortunately, would not be enough to lessen the overall numbers staying in the valley. Another suggestion was to place a limit on the number of people allowed in the valley, closing it once that number had been reached. But such a policy would be difficult to manage and would be greatly unpopular, leading to a public backlash. Finally, they suggested limiting the length of visitors' stays. Such a policy appeared to be the best possible solution, and the public, they argued, viewed such policies as fair.[57]

The Park Service largely agreed with the Board's conclusions. Having guided Yosemite through the war, Superintendent Kittridge worried that the continued growth of tourist facilities made the Yosemite Valley seem more like a resort than a national park. Of particular concern was the expansion of Camp Curry, which no longer embodied the "delightful, homey, and wholesome" appeal of its past. Instead, it now resembled a village with a carnival atmosphere, full of town amusements. The camp now included a "huge dance hall" and lodging for roughly 1,500 a night, he lamented, in a location ill-suited for such numbers. To preserve the valley's scenery and character, Kittridge argued for the decentralization of campgrounds and other facilities throughout the park, as well as removal of all administrative buildings to outside of the park. Doing so, he believed, would eliminate the "jazz or city type of amusements" within Yosemite, which attracted many to the park.[58]

Understandably, the YP&CC disagreed with Kittridge's view, countering that visitors had come to expect such amenities. These two differing views set the stage for the coming decade's battle over Yosemite's purpose, and the public's paradoxical desires for greater amenities and preservation of the Yosemite they knew.

4 Let's Open the Parks

Bernard DeVoto was furious, and he wanted his readers to know why. The renowned author, historian, and social critic was still waiting for a promised delivery of a brand-new 1946 station wagon from whom he identified as a "great automobile manufacturer." He had been assured, when he had paid his deposit months ago, that he would receive one of the first vehicles off the assembly line. But he still did not have his new car. In his monthly "Easy Chair" column for *Harper's* magazine, DeVoto grumbled, "At this moment that station wagon is splashed in gorgeous colors across the advertising pages of the United States; I see it glistening in every magazine I pick up. There are no station wagons, but there are inspiring ads." DeVoto had agreed to purchase the car for his family's tour of the American West that coming summer, during which he would write about the war's impacts on the region.[1]

The DeVoto family had planned the road trip four years earlier, but the United States' entry into World War II had dashed those plans. Now, with Japan's surrender, the promise of the open road beckoned once again. DeVoto had agreed to chronicle the family's travels for several newspapers and magazines in order to fund the three-month vacation. He had even secured sponsorship from the "great automobile manufacturer." But it was now clear to DeVoto that no such car would be arriving in front of his Cambridge, Massachusetts, home anytime soon. So, rather than wait, the family decided to pile into their 1941 Buick sedan and head west.[2]

The family traveled first to Michigan, then on to Wisconsin, Minnesota, and South Dakota before reaching Montana, where DeVoto filed his first account of the trip for *Harpers*. Rather than an adventurous tale of the open road, he instead delivered a litany of complaints about the quality of food, lodgings, and roads they had encountered on the trip thus far. DeVoto continued to voice his grievances in his following month's column, noting that the war had done no favors to the country's tourist stops, local cafés, and highways. It was not all bad news, however. The food in Montana and Utah was particularly good, if somewhat simple. And, thankfully, the Buick was holding up well, having required only a few minor repairs along the way.

By July the family had reached Portland, Oregon. From there, after celebrating the Fourth, they continued south, first to Crater Lake, then to San Francisco before making their way to Yosemite. The family spent only a few short hours in the park before heading east to Nevada and then on to Idaho. After dropping their eldest off at a ranch, where he worked for the remainder of the summer, and their youngest at a youth camp, Bernard and Avis crisscrossed the intermountain West over the next two months. Driving long distances, they visited sites Bernard had written about but had never seen, including several national parks. By the time the family returned to Cambridge, they had covered more than 13,000 miles, a point DeVoto took great pride in telling his readers. He had published three "Easy Chair" columns chronicling the trip throughout the summer. In each, his running commentary on the West's lack of decent lodgings and passable meals and his adventures on roads too rough to enjoy was countered only by his reverence for the region's raw natural beauty.[3]

Yet, of the thousands of words he wrote that summer extolling the West's splendor, DeVoto spent only thirty words on Yosemite, a passing mention of his hair-raising trip on the Tioga Road. That the view from Inspiration Point, the grandeur of Half Dome, and the beauty of Tuolumne Meadows failed to capture the journalist's attention, or even merit a mention, underscored the Tioga Road's poor condition. Specifically, a twenty-one-mile section of dirt road that the Park Service had left unimproved before the war because of the rugged terrain it traversed. Notoriously narrow and winding, the section required drivers to navigate steep, often washed-out sections, impossibly tight switchbacks, and other drivers. Traffic along the road backed up for hours as cars overheated, trailers jackknifed, and tempers flared. Those who dared the drive found it exasperating, one driver noting dryly, "It is a good deal like a roller coaster, only rougher." As visitors soon discovered, the Tioga Road's condition was emblematic of the park's larger disorder.[4]

Visitation exploded from 290,000 visitors in 1945 to more than 820,000 just five years later. Within a decade, visitor numbers crested the one million mark for the first time in Yosemite's history and showed no signs of letting up. The park's infrastructure proved woefully inadequate to meet such numbers. Campgrounds lay in disrepair throughout the park. Similarly, much of the Yosemite Village was run-down, while traffic was far too often bumper to bumper as it crawled through the valley. In an effort to manage the growing number of cars, the Park Service razed the Curry orchard to build a parking lot, but even that failed to meet the need for more parking.

In addition, finding a camping spot at the height of the summer season often proved an exercise in endurance with more than 12,000 visitors seeking to fit into camping spots built to accommodate 4,000 people. Worries that such crowds were destroying the very thing that made Yosemite special grew as fast as visitor numbers. As Yosemite's superintendent, Carl Russell, ominously observed, "Unless something is done to stop present trends much of the natural appeal of the valley as we have known it will be gone in another fifty years." Such apprehensions led to countless proposals on how to best resolve overcrowding within the park.[5]

Some suggested opening the entire valley floor to camping. Doing so, they reasoned, would spread campers out, making their stay more enjoyable. One of the early proponents of this idea was *Los Angeles Times* columnist Martin Litton, who, in 1952, argued that Yosemite should not serve as a recreational playground with amenities like a golf course, but should instead be preserved as a sanctuary where visitors could appreciate nature's scenic wonders. Others proposed the removal of all lodges to the park's periphery, leaving the valley uninhabited except for day visitors. Many saw banning automobiles from the Yosemite Valley and replacing them with buses that would transport visitors between points of interest as the only viable solution. To that end the Park Service maintained that recreation remained the most suitable and democratic use of the park, a perspective reflected in its 1952 Master Plan.[6]

Like previous plans, the new master plan sought to balance visitor use with protecting Yosemite's scenery for visitors' continued enjoyment. To accomplish that goal, the Park Service halted the expansion of the Glacier Point Hotel and Big Trees Lodges, asking instead to develop more lodging at Wawona and Tuolumne Meadows. Moreover, the agency planned on renovating all campgrounds within the Yosemite Valley, as well as building new campsites in Wawona, Crane Flat, and Bridalveil Meadows, in efforts to spread overnight stays more equitably across the park. The goal was not to limit the number of campers in Yosemite, but to disperse them throughout the park.[7]

How to pay for these new campgrounds and facilities proved to be a significant stumbling block, however. Limited budgets and staffing had made maintenance in Yosemite nearly impossible during the war. Congress's rebuffing any spending that smacked of New Deal programs, combined with the escalating costs of the Korean War, further reduced the Park Service's budget, worsening Yosemite's state of disrepair. It was not only Yosemite that struggled with limited budgets and growing issues. As DeVoto had experienced, every national park was underfunded and in poor shape.

As his biographer Wallace Stegner noted, DeVoto left Cambridge in 1946 a tourist but returned a conservationist. The journalist had grown increasingly horror-struck throughout the family's Western journey over the dismal state of the nation's public lands. Grasslands were overgrazed, and the national parks lay in an offensive state of shabbiness. Upon returning to Massachusetts, the cantankerous DeVoto laid siege to those he considered most responsible for the damage being wrought on the public's domain, starting with Western cattle and sheep ranchers. Their greed and carelessness had destroyed much of the West's grazing lands, he raged. While Western cattlemen's associations rejected DeVoto's attacks as the naive rantings of an eastern academic elite, DeVoto's readers came to share his views, taking them on as their own.[8]

But it was for the deteriorating state of the national parks that DeVoto saved his greatest vitriol. A lack of adequate congressional funding, he argued, was causing the parks to fall into disrepair. Wildfires were being allowed to burn unchecked, bridges in some parks were so neglected that they had been condemned, and staffing in most parks remained sparse at best. Yet DeVoto did not blame the Park Service for this neglect, avowing, "No bureau is more efficient, more dedicated, more hard-working, or public-spirited than the Park Service." No, it was not the Park Service that was failing the American public, he argued, but Congress's refusal to adequately fund the agency to staff, maintain, and, yes, even develop needed facilities that had allowed the national parks to reach this point of crisis.[9]

Members of Congress largely agreed with DeVoto's assessment that funding had not kept pace with demand. Amid postwar economic uncertainty, the costs of the Korean War, and political pressures to maintain a balanced budget, the upkeep of the national parks remained a relatively low priority. Despite this, Americans continued to flock to their national parks in record numbers, exacerbating the strain on an already overstressed system.

At wit's end with Congress's inaction, DeVoto penned one of his most celebrated "Easy Chair" polemics, "Let's Close the National Parks." In it, he condemned Congress's unwillingness to properly fund the parks and demanded that if the government was unwilling to spend the necessary money to maintain the parks then they should simply close them. In fact, he had a list of which parks should be shuttered first. "Let us, as a beginning, close Yellowstone, Yosemite, Rocky Mountain, and Grand Canyon National Parks—close and seal them, assign the Army to patrol them, and so hold them secure till they can reopen," DeVoto thundered. Such a disgrace, he argued, would finally gain the attention of those responsible for the ruin of the national parks, the U.S. Congress.[10]

DeVoto's editorial struck a deep chord with his readers, who, like DeVoto, had grown impatient with the lack of upkeep, camping spots, and other amenities within their national parks. The Park Service's new director, Conrad Wirth, openly acknowledged that due to an inadequate budget the agency could not provide essential services, including sanitation and visitor safety. Campgrounds in many parks looked more like rural slums than national parks, he warned, and the water, sewer, and electrical infrastructure in most parks were stretched beyond their limits. "We actually get scared when we think of the bad health conditions," Wirth confessed in an interview with *Reader's Digest*. Wirth's admission reinforced the growing account that the nation's parks were crumbling. Following DeVoto's jeremiad, newspapers and magazines across the country ran their own stories condemning the state of the parks. To meet these challenges, in 1953 Wirth proposed an ambitious ten-year plan that would infuse a billion dollars into the parks. Named Mission 66, the program would forever reshape the national park system and led to a furious debate over Yosemite's popularity.[11]

BUILT ON STEPHEN Mather's philosophy that development was the surest way to preserve national parks and that the parks were the best way to uphold American values, the decade-long Mission 66 program invested millions of dollars in modernizing the national park system. While presented to the public as a new initiative, the program was largely a response to the growing demands that the federal government do something to restore the national parks to their previous state. This meant building new visitor centers, upgrading trails and roads, improving sanitation, and hiring more rangers to meet growing visitation across the national park system.[12]

While this was a welcome relief to much of the touring public, not all agreed on the program's goals. A growing number of environmentalist voices opposed both the aesthetics and the scale of the new construction within the national park system. Many argued that Mission 66's focus on development was at odds with what they believed should be the Park Service's mandate: ecological rehabilitation and preservation. How could the Park Service claim to be preserving parks like Yosemite while allowing the construction of new visitor centers, hotels, campgrounds, and even roads, a growing number of environmentalists asked. Shouldn't the agency's primary mission be the conservation of natural resources rather than the continued development of parks for tourists' enjoyment? The Park Service responded to such questions that Mission 66 provided much-needed visitor facilities and programs while still fulfilling its obligations under the Organic Act to promote and regulate the national parks for visitors' enjoyment.

In Yosemite, this meant protecting the Yosemite Valley by relocating as many administrative services as possible to outside of the valley and holding the line on expansion of any further accommodations within the valley. Road and trail systems were to be completed, primarily along the Crane Flat and Tioga Road entrance routes, to relieve pressure on the South and Arch Rock entrances and reduce traffic congestion within the valley. New water and sewer systems would modernize the park's antiquated sanitation facilities. Obsolete concession buildings would be replaced or upgraded, and additional accommodations would be constructed to relieve overcrowding. To reduce overcrowding within the valley, the Park Service added hundreds of new campsites at Bridalveil Creek and Crane Flat and built new visitor centers at Wawona, Big Meadow, and El Portal. Seven new interpretive rangers would be hired, and a new museum on Glacier Point would be built celebrating the valley's natural history. Finally, the few private inholdings that still existed within the park would be acquired on the principle that they conflicted with the goal of maximizing public access.[13]

Among its numerous goals, the Park Service prioritized completing the middle twenty-one-mile stretch of Tioga Road. Built in 1883 by the Tioga Mine Company to transport its silver ore from the Sierra Nevada high country down to its smelters in the Central Valley, the narrow wagon route had been abandoned nearly immediately after its completion due to plummeting silver prices. Travel along the road remained sparse over the next several decades, until early motorists rediscovered the route at the beginning of the twentieth century. A point not lost on Stephen Mather, who as early as 1915 saw the road as a means to open Yosemite's high country to visitors. A fervent national park booster, the newly appointed assistant to the secretary of the interior had made millions as the owner of the Thorkildsen Mather Borax Company. With the federal government unable, and unwilling, to purchase the road, Mather offered to open his own checkbook to purchase the road for the Park Service. But when told his plan ran afoul of federal ethics laws, Mather turned to wealthy benefactors to buy and then donate the road to the Park Service. Constructed for stagecoaches, however, the narrow, winding, narrow track proved far too inadequate to accommodate modern automobiles, compelling Park Service landscape architects to redesign the road.[14]

Dividing the road into thirds, architects focused first on widening its eastern and western segments by twenty feet and lowering all grades to a maximum of 6 percent. After completing these two sections in 1938, the Park Service turned its attention to the road's much more rugged middle stretch. The twenty-one miles of steep and winding dirt road traveled through some of Yosemite's most spectacular country, including along the shores of Lake Tenaya and over vast stretches of glacially polished granite. But what made the section so stunning also made it difficult to build a modern road through. Particularly the exposed granite above Tenaya Lake, which would require blasting the road through tons of rock in order to comply with modern standards.

As early as 1933, state engineers had considered an alternate High Line route north through the Ten Lakes region avoiding the granite section altogether. But the Park Service resisted that option, believing the longer route would unnecessarily invade the park's backcountry, and would take visitors away from the stunning views along the established route. After conducting a preliminary report on the two options, the Park Service, along with the Yosemite Advisory Board and the Sierra Club, settled on the lower route along the lake as the best option. But the nation's entry into World War II halted all construction, and the road remained unfinished.[15]

For inexperienced mountain drivers, the unfinished section of the road proved to be more than they had bargained for. As the American Automobile Association noted the year after DeVoto drove through the park, "It is not unusual to find people unused to mountain roads who just go to pieces, freeze at the wheel and park the cars in the middle of the road and wait for the Park Rangers or a kindly motorist to drive their cars the rest of the way." The road tested not only drivers' skills but their patience. Minor accidents often left cars with dented fenders and bumpers. Those foolish enough to tow a trailer over the Tioga Road could cause traffic jams that lasted for hours. Tempers flared when drivers unaccustomed to narrow mountain roads found themselves crawling along at twenty miles an hour. Despite such circumstances, by 1950 traffic along the road had increased by 30 percent over prewar levels.[16]

Due to such realities, by the mid-1940s the Park Service's chief landscape architect, Thomas Vint, argued the road was no longer safe and was in dire need of improvement if the Park Service sought to fulfill its obligation to make the park accessible. But while supporting the widening and straightening of the twenty-one-mile section of road, Vint warned that any construction could gravely affect the region's scenery, the preservation of which remained "highly important." Over the coming decades, balancing these two seemingly disparate points shaped an emergent debate over Yosemite's, and, more broadly, the national parks' fundamental purpose.[17]

AS THE PARK Service resolved to improve the Tioga Road in order to increase its safety and further access to Yosemite, two men with deep ties to Yosemite became the faces of the opposition to further development within Yosemite by calling for the preservation of the park's wild places. Through their battles with the National Park Service over the Tioga, Ansel Adams and David Brower helped redefine American environmental politics and the core idea of the national parks in the 1960s.

For Adams the clash over the Tioga Road had begun nearly three decades earlier, when, after a two-day trudge from the Yosemite Valley, he and two friends arrived on Lake Tenaya's shore and made camp along the banks of a small stream that emptied into the crystal clear lake. Late that evening, with a cool wind blowing wisps of smoke from a fishing camp's campfire across the lake, the smell of frying fish and wood smoke filling the air, Adams wandered to the lake's shore. Standing silently alone, the twenty-year-old pianist and budding photographer contemplated the moonlight gleaming off the surrounding granite slopes, the distant sound of waterfalls, and a nearly hushed creaking of the pine trees swaying in the breeze. "The meaning of nature is resonant and clear in such experiences," Adams wrote of the moment nearly thirty years later. "These are the symbols of a reality beyond the material works of man and beyond the ability of his art to express fully." Such experiences could not be gained by the swift whirr of wheels on pavement, Adams argued, only by the crunch of dusty boots on the earth.[18]

Adams's relationship with Yosemite spanned much of the twentieth century. Raised in San Francisco, he first visited Yosemite Valley in 1916 with his parents. It was love at first sight. His father had given him a Kodak No. 1 Box Brownie Camera for the trip. The excited fourteen-year-old spent the next several days racing around the valley, snapping photographs. While his snapshots were typical of many tourists and not yet gallery-worthy, the die had been cast. Three years later Adams returned to Yosemite to work as the caretaker of the Sierra Club's LeConte Memorial Lodge, a position he had secured through his family friend Francis Holman. A retired mining engineer and Sierra Club member, Holman introduced the young Adams to the High Sierra. They spent two days exploring Yosemite's high country, sleeping under the stars and summiting remote peaks. Adams continued to visit Yosemite every summer over the next several years, honing his skills as a landscape photographer by trekking throughout the park with his camera in tow.[19]

During those years Adams met, fell in love with, got engaged to, broke up with, got engaged to again, and finally married Virginia Best. The daughter of painter Harry Cassie Best, who owned a studio in Yosemite Valley that she would one day inherit, Virginia was among the first to summit the Kaweah Peaks in Sequoia National Park and was the first woman elected to the Sierra Club's board. The couple operated the small Best studio for several decades, making a living by selling Adams's photographs to visitors. Virginia remained an ardent environmentalist throughout her life, collaborating with her husband on several projects, including a children's book about her childhood in the valley, and quietly running the family business.

Adams's breakthrough as a photographer occurred in 1927, when having spent the day hiking and taking photographs with Virginia and their group of friends, Adams arrived at the Diving Board, a granite outcropping on the west side of Half Dome, with only two photographic plates left with which to shoot the face of Half Dome. As soon as he pressed the shutter on the first plate, he knew he had missed something, and he quickly swapped the yellow filter for a deeper red filter. The result was breathtaking. The filter had darkened the sky to nearly black, highlighting Half Dome's stark granite face. Titled *Monolith, the Face of Half Dome, Yosemite National Park*, the image set Adams on a course to becoming one of the era's preeminent landscape photographers.[20]

Over the next four decades, Adams produced some of the most iconic landscape photographs in modern history. Often shot in black and white, his images were at once nostalgic and transformative, capturing both sweeping vistas and intimate portraits of seemingly untouched nature. As the Museum of Modern Arts director John Szarkowski later asserted, Adams's photographs were anachronisms, presenting natural landscapes as somehow devoid of human industry. By framing nature in such a manner, Adams evoked artwork by painters like Albert Bierstadt and photographers like Carleton Watkins, while offering a stark modernized portrayal of Western landscapes. In a 1951 profile of Adams, *Time* magazine noted his photographs, especially those of his beloved Yosemite, inspired millions to follow him into "the National Parks fixing the cold eye of their cameras on the same splendors he has photographed—and hoping, somehow, to match his art." But on arriving many found Yosemite overrun with others seeking that same connection, tarnishing the romantic vision Adams promised through the lens of his camera.[21]

His pictures held the promise that Yosemite could still be experienced far away from the noise and the crowd—a vision grounded in Adams's own experiences as a young man exploring Yosemite's backcountry. Building a modern highway through Yosemite would not only destroy Yosemite's wilderness, he argued, but deprive future generations of experiencing it as he had all those decades ago.

A decade younger than Adams, David Brower also maintained a lifelong connection to Yosemite. Like Adams he first visited the park as a child during a family vacation. Traveling from their home in Berkeley, the Browers loaded camping gear onto the running board of their 1918 Maxwell and headed east into the mountains, stopping first in Sacramento then proceeding to the small town of Colfax, over Donner Pass, and to the shores of Lake Tahoe before heading south toward Mono Lake and the eastern entrance to the park. They then drove over the infamously steep Tioga Pass Road. At one point young David, his brother, and his mother had to jump out and push the family car around a switchback. The family eventually reached the Yosemite Valley, where they spent the next several days exploring.[22]

The trip left a lasting impression on Brower. A bright student, he had skipped several grades before enrolling at the University of California. But school could not hold his attention as much as the Sierras and rock climbing. At age twenty-one, Brower dropped out of the university and turned his attention to mountaineering. It was an auspicious decision. Over the next decade, he made a name for himself as a daring climber, becoming the first to scale numerous Sierra peaks, including seventy first ascents within Yosemite. On the suggestion of a friend, Brower joined the Sierra Club in 1933. Two years later he found work as a cashier with the Yosemite Park & Curry Company. It was a dream job, allowing him to stay and climb in the park throughout the entire summer. He soon began selling stories of his climbs to newspapers and magazines, establishing himself as one of the nation's renowned mountaineers. But World War II interrupted Brower's charmed life in the Sierra Nevada. Heeding his country's call, he enlisted in the army as an infantryman and eventually served as a lieutenant in the Tenth Mountain Division. The storied unit included some of the best mountaineers and skiers in the nation and helped launch both the modern ski industry and the environmental movement.[23]

While serving in the Italian Alps, Brower grew increasingly alarmed at the destruction of Europe's alpine environment, and he returned to Berkeley, following the war, focused on saving his Sierras from a similar fate. After a brief stint with the University of California Press, he took a job as the editor of the *Sierra Club Bulletin*, and he eventually became the organization's executive director. Together, Brower and Adams represented a generational and political shift within the Sierra Club. Nostalgic for the Yosemite of their youth, the two men had grown disillusioned over the Park Service's favoring visitor access over the preservation of nature. Nothing symbolized the Park Service's failure to preserve Yosemite's natural wonder more than its planned expansion of the Tioga Road.

Like Adams, Brower viewed the Park Service's road widening efforts as contrary to the agency's mandate and indicative of the increasing threat that recreation posed to the environment. In 1949, just two years after the Advisory Board and the Sierra Club had agreed to blast the glacial granite above Lake Tenaya, Brower and fellow Club member Harold Bradley co-authored a scathing critique of the Park Service's decision. Their thirty-three-page missive condemned the agency for prioritizing accessibility over preservation and criticized the construction of all new roads within national parks. Pointing to the overcrowding in Yosemite Valley, the future Sierra Club president observed that "a visit to Yosemite today has little of the pilgrimage element it once had" and warned that continued highway construction into wilderness areas would lead to the park's inevitable deterioration by opening the Tenaya Lake region to further development.[24]

The struggle over the Tioga Road also highlighted a growing schism within the Sierra Club. Older members, like executive director William Colby, supported the road's expansion, arguing that improving the Tioga Road aligned with the Club's long-held mission of making the Sierra accessible for all to enjoy. In contrast, younger members criticized this view as overly simplistic and detrimental to the very places the Club aimed to protect. Frustrated by the increasing resistance from within the organization, Colby resigned as its director in 1946. Although he remained active with the Club until his death in 1964, his resignation marked a significant shift in both the Sierra Club's leadership and broader American environmental politics. The preservation of wilderness came to increasingly dominate not only the Sierra Club's agenda but the national dialogue over the state of the environment.

This debate came to pit preservationists like Brower, Adams, and Bradley against longtime allies within the Park Service, changing the very dynamic that had shaped the parks for the past half century.[25]

Bradley returned to the theme of automobiles and overcrowding in 1951. Writing in the magazine *Pacific Discovery*, he argued that opening Yosemite's alpine country to unlimited traffic would be akin to removing the "QUIET" sign at a library entrance. Soon, visitors would demand that the road be plowed in winter and call for the development of ski resorts, ultimately forcing those who sought a peaceful retreat into the wilderness to find somewhere else, he argued.

Bradley saved his bitterest vitriol for camper trailers, attacking "those who require a house on wheels when they rough it" as too timid, or incompetent, to enjoy the road's rugged beauty. Bradley declared trailers had no place within Yosemite's high country. Others shared Bradley's sentiments, but for different reasons. Many visitors blamed trailers for the park's growing traffic woes. Others complained about how trailers crowded campgrounds. Despite such condemnations, camping trailers continued to grow in popularity. Complete with kitchens, dining tables, and beds, recreational trailers and other vehicles allowed Americans to bring all the conveniences of home on the road. But as many owners found, although they could bring their home away from home with them to Yosemite, there was no guarantee the park would be open to them.[26]

DRIVING FROM THE small desert community of Lancaster, California, the McArthur family had had a rough time during their previous visit to Yosemite. First, the couple had failed to secure a camping site within the Yosemite Valley that was level enough to park their trailer. Then, in Wawona, eight men had to help push their trailer back onto the highway after it became stuck in the only available parking spot. To avoid similar issues on their next trip, Alexander McArthur wrote a letter to Yosemite's chief ranger inquiring about the availability of electrical hookups for trailers, washing machines for laundry, butane, and parking spaces. In a brief response, a park ranger explained that there were no electrical or sewer connections in any of the national park's campgrounds, and that the park did not offer any special campsites for camping trailers. However, there was a laundromat located within the valley.[27]

The McArthurs were not alone in their struggle to find an adequate spot to park their trailer within Yosemite. While visitors had been bringing camping trailers of one sort or another into Yosemite since the 1920s, the manufacture of larger, more lavish trailers, which included full kitchens, toilets, and even showers, was quickly changing how Americans enjoyed the outdoors. In 1955 alone over 8,000 towed their trailers into Yosemite looking to enjoy the great outdoors with all the amenities of home. Within two years that number rose to just under 10,000. While such numbers represented only a fraction of overnight stays within Yosemite, they represented mounting expectations among visitors that the Park Service accommodate their needs, including larger sites complete with electricity, sewage, and water hookups, along with the traditional picnic tables, firepits, and nearby restrooms.[28]

Fred Grey was among those expecting such services in 1955. And he was even willing to pay a nominal fee for an electric hookup in one of Yosemite's numerous free campgrounds. One of the thousands of Californians to tow their trailers into the park, the Inglewood resident had grown accustomed to campgrounds offering such amenities as electricity and sewer. After finding Yosemite's campgrounds lacking even what he considered the simplest of services, Grey took up pen and paper and suggested that the Park Service could better serve visitors by offering such modern conveniences.

Acting Superintendent Chas Hill responded by thanking Grey for his suggestion but explaining that the current power lines within the park were already overloaded and allowing the few hundred trailers electrical hookups would be nigh impossible, particularly when trailers made up just a fraction of the nearly 11,000 visitors staying in one of the park's free campgrounds at any time during the busy summer months. "We appreciate your sincere interest in improving service to the camper in Yosemite," Hill wrote to Grey, "but at this time due to a lack of funds we can only maintain present facilities." Yosemite was a national park after all, not a commercial campground.[29]

Making matters more difficult, many lacked experience in pulling what amounted to a small house behind their family sedan. Such inexperience contributed to traffic jams, crowded campgrounds, and tensions among tourists competing for parking. Furthermore, the comfort these trailers afforded made it all too easy for visitors to ignore the park's fifteen-day summer camping limit. Yet the reality of the park had not yet overtaken its mystique. Thousands continued to bring their trailers to Yosemite looking to enjoy the park, roasting marshmallows and hotdogs over an open campfire, watching the Firefall, and sleeping in their trailer.

Among them was the Williams family. After driving through multiple campgrounds, the family of four finally happened upon the perfect site.

Located on the banks of the Merced River, with a broad gravel beach for the kids to play on, and a wide enough spot among the trees to park their borrowed tent trailer, the spot would serve as the family's home for the next two weeks. After they'd waited two hours for the former occupant to load up and pull out, it was theirs. "Now we'll see how much you really know about camping," Geegee teased her husband, Dick, as the kids raced to the river to play.

Married for nine years, the couple had met at a party during the war, fallen in love, and eloped. A justice of the peace presided over a private ceremony in Yuma, Arizona. After returning to California, they spent two days together in San Francisco before Dick shipped out to the Pacific, and Geegee, short for Virginia, returned to her job riveting gas tanks for Douglas Aircraft. After Dick returned from the war, the couple moved in with his parents, had their first child, and built a house on a small lot they had bought in the beach community of Venice. Dick soon found work as a firefighter, while Geegee stayed home to raise their now two children.[30]

The Williamses were among the nearly half a million visitors making the six-hour drive from Los Angeles to Yosemite that summer. By the early 1950s, Californians made up roughly two-thirds of all visitors entering Yosemite. And of those, most hailed from Southern California's flourishing suburban empire. That the majority of Yosemite's visitors came from California's burgeoning metropolitan areas was no surprise. For generations, those living in the San Francisco Bay Area had considered Yosemite as their backyard. Many still do. Yet the sheer number of Southern Californians streaming into the park reflected Los Angeles's growing significance within the state. The move of the Brooklyn Dodgers to Chavez Ravine, along with the opening of Disneyland in Anaheim, underscored the region's postwar leisure boom. And just a mere day's drive away, Yosemite offered young families like the Williamses an affordable nearby vacation destination.[31]

Over the next week, the family fell into a routine. Waking with the sun, Dick, who never cooked at home, prepared breakfast. After shoveling down a healthy pile of pancakes, the family cleaned up and headed out to sightsee. They climbed to the base of Vernal Falls, made the hour drive to the Mariposa Grove, where they drove through Tunnel Tree, and stood in awe of the view of the Yosemite Valley from Glacier Point. Sated with the sights, the Williamses returned to their campsite, and while Dick and the children played in the river, Geegee either napped or made her way to one of the valley's four general stores to grocery shop or do laundry. "Pretty civilized camping," she noted of the park's modern amenities, "but a big help when you go with children." Evenings centered around a dinner cooked over the open flame of the campfire, cleaning up, and wandering over to Camp Curry to watch the Firefall. Exhausted, the four returned to their trailer and fell blissfully asleep until the next morning when they awoke and started again.[32]

But it was not all sightseeing and shopping. Their first night in the park, Geegee woke to a crash outside the trailer. Alarmed, she woke Dick, who had packed the family's food in a duffle and hung it from a tree to keep it from bears. Unfortunately, he had not hung it high enough, and one of the valley's resident bears had discovered the family's cache. Poking his head outside, Dick saw the black bear messily nosing through the bag. Terrified, Geegee pulled him back in the tent and listened as the invader ate their food. "She didn't know I couldn't walk," Dick later confessed retelling the tale. Finally, a neighboring camper chased off the invader armed with a wooden spoon, pounding on a dishpan and yelling, "Where's that bear, where's that bear!"

The highlight of the vacation, though, was the family's overnight trip to Lake Tenaya. Leaving the trailer behind, the Williamses made their way out of the valley to the Tioga Road. Along the way Dick helped a fellow visitor turn their trailer around manually because of the road's steep grade. Later, they sat and waited for twenty minutes to allow cars to come from the opposite direction through one of the narrow sections. But the journey proved to be worth the headaches once they arrived at Lake Tenaya. Geegee could not believe the lake's beauty. "You can't believe it's real," she recalled, on first seeing the granite peaks mirrored in the lake's crystal clear waters.

On arriving, a ranger showed them to a designated spot with a campfire ring only a few feet from the shore. That evening the family enjoyed a dinner of canned goods cooked over the open flame after which Geegee realized that she had forgotten to bring a dishpan to wash the dishes. So instead she scraped the dishes in the sand, burying the refuse on the beach and rinsing the plates and silverware in the lake. After putting the children to bed in the car, Dick and Geegee wrapped a blanket around themselves and enjoyed the campfire, the wood for which Dick and little Dickie had scrounged from the nearby trees. That night proved to be colder than the Williamses had expected, and the rays of morning sunshine were a welcome sight. After two weeks in Yosemite, Williams reluctantly packed up and headed back to Venice. On the drive home, Dick mused aloud about the trip. It had been a great vacation. The first, in fact, the family had ever taken, and something he hoped to do again. It was a hope shared by millions of other Californians.[33]

By the early 1950s, America's emerging middle class embraced the family vacation with gusto. Loading the children into the family car and heading to a national park became a part of the American experience. Yet as families like the Williamses flocked to Yosemite seeking respite, they brought the suburban landscapes they had left behind with them. Weekend traffic in the park increasingly felt more like morning rush hour than a leisurely drive through the woods, and the general store carried everything from fresh fruit to blue jeans. Set free in this wonderland, children reenacted the exploits of their favorite television heroes, crawling over rocks and racing through the trees. A heady blend of consumerism and sublime wonder, such experiences formed Americans' connections with the national parks, and understandings of the natural world.[34]

However, the completion of the New Village in 1959, along with the construction of the New Yosemite Lodge three years earlier, fueled growing criticisms that the NPS prioritized recreational development above preservation. While the Sierra Club and the NPS battled over the widening of the Tioga Road, another bitter debate, over the environmental impacts of climbing, played out on the Yosemite Valley's granite walls.[35]

FOR GENERATIONS ADVENTURERS had sought to conquer the Sierra Nevada's granite peaks. In 1865 members of the California Geological Survey, led by Clarence King, summited what would later become known as Mount Whitney. John Muir's scramble to the top of Cathedral Peak four years later, followed by George Anderson's ascent of Half Dome in 1875, were among the first feats of mountaineering within the Yosemite Valley. Yet many of Yosemite's granite monoliths and towering spires remained out of mountaineers' reach until the introduction of European pitons to North America in the 1930s. A simple steel spike with a hole on one end, which allowed a rope through that was then tied around the waist of a climber, pitons allowed climbers to ascend ever-steeper rock faces. Either pounded into natural cracks or hammered into hand-drilled holes, the new technology revolutionized climbing. Within a decade the sport ballooned in popularity from one end of California to the other, drawing men and women to test their mettle by scaling cliff faces and boulder fields throughout the state. This was particularly true within Yosemite.

Climbing in Yosemite entered what many historians of the sport consider its golden age, in the years following World War II. In 1946 climbers John Salathe and Anton "Ax" Nelson completed the first ascent up the southwest face of Half Dome. The duo's feat forever changed climbing not only in Yosemite but around the world. Born in Switzerland, where he had trained as a blacksmith, Salathe played a critical role in Yosemite's emergence as climbing's Mecca. He had spent much of his youth rambling around the globe before finally migrating to Canada. There, Salathe married and started a family. But wanderlust struck again, and the family moved to San Mateo, California, where Salathe opened the Peninsula Wrought Iron Works. But the California dream they hoped to acquire was not to be.

By the end of World War II, both Salathe's marriage and health were failing. Seeking solace, he discovered rock climbing by joining the Sierra Club outings to Yosemite. Ever the tinkerer, Salathe quickly became disenchanted with the pitons designed for climbing in the Alps and began forging steel pitons more suitable for the Sierra's harder granite in his home forge. Armed with his stronger pitons, over the next two years Salathe accomplished first ascents up not only Half Dome's southwestern face but Lost Arrow Spire and Sentinel Rock. Such feats proved that more was possible, and a new generation of climbers set their sights on Yosemite's big walls.[36]

Yosemite's era of big wall climbing began on a warm June morning in 1955. Having spent the night at the base of Half Dome, Royal Robbins, Jerry Gallwas, Don Wilson, and Warren Harding began climbing. The four men had met while climbing over the previous two summers before deciding to attempt the ascent up Half Dome's north face. Their efforts ended in failure after only two and a half days. Having climbed only a third of the way, and running short on food and water, Robbins, Wilson, and Gallwas agreed to quit. Harding disagreed, arguing that the group should persist and finish the climb. The rest of the team overruled the boisterous Harding, and the four retreated to the valley floor.[37]

While they failed to reach the summit, the men's attempt showed that scaling the wall was indeed possible. Almost perfectly vertical, the cliff face's sheer vastness proved to be the primary challenge in completing the climb. The effort also exposed philosophical differences between the hard-charging Harding and the exacting Robbins—a difference that played out over the next four years, defining both big wall climbing and the Yosemite Valley. Robbins and Gallwas, along with Mike Sherrick, returned two years later to complete the climb. Arriving a few days late to begin his ascent, Harding hiked to the top of Half Dome and met the three men as they crested the summit. But not being the first to conquer the vertical face of Half Dome had always been an issue for Harding, and he spent the next several weeks wandering around the valley seeking some way he could surpass the achievement.[38]

Only one wall remained unconquered within the valley, El Capitan. Climbers had long viewed the nearly vertical face with a mixture of awe and fear. Only one route appeared possible up the precipice, a line Yosemite Valley climbers referred to as the Nose. But such a climb would require a tremendous amount of rope, pitons, and bolts, and months of effort. Even then, many considered it impossible. Rising to the challenge, Harding recruited fellow climbers Will Feuerer and Mark Powell to lay siege to the three-thousand-foot wall. And so in July of 1957 the three men began their ascent up the Nose.

To Robbins, Harding's use of bolts hammered into the granite surface was an affront to the sport's aesthetic, and he attacked the climb as an anathema. The men's rivalry extended beyond their philosophy on climbing. Quiet-mannered and bookish, Robbins's personality differed almost entirely from the bombastic Harding. The two men's temperaments reflected how they climbed. Harding charged any climb with gusto, simply wishing to conquer the most challenging climbs in the valley by any means possible, even if it meant drilling hundreds of holes into the rock to secure bolts. While Robbins also sought to conquer Yosemite's sheer granite faces, he eschewed bolts, arguing they took away the adventure of challenging yourself against nature. Furthermore, drilling holes into a face of granite forever scarred the rock climbers relied upon. Harding was untroubled by such moralistic distinctions, arguing bolts were a necessary means to climb where no natural hold existed. By the end of the 1950s, these two philosophies, and the men who championed them, clashed over not just the ethics of climbing but a much larger debate over the use of technology and the environment.[39]

But the ascent was causing more than a fight among the valley's climbing community. After the three men climbed a third of the way up the Nose, crowds of spectators began to gather in El Capitan meadow to watch their ascent. "It seemed that our climbing presented quite a spectacle and had attracted a crowd of tourists, which created a traffic jam at the road-junction near the base," Harding recalled upon his return to the valley floor. Realizing the curiosity they had become, the three climbers agreed with the Park Service to postpone their next climb until after the Labor Day weekend, when, theoretically, fewer visitors would gather below.

Over the ensuing months, the trio held several conversations with Park Service officials, leading to an agreement in October that the climb would proceed under the stipulations that the climbers would use bolts and ropes sparingly in their ascent, that the climbing party would include enough members to meet any emergency, that the climb would be complete by Memorial Day, and that if the Park Service determined the climb was attracting an "audience of spectators detrimental to the best interest of the National Park Service," then it would be postponed. But before signing off on the terms, Harding penciled in one last stipulation of his own. "We understand that if the climb is making satisfactory progress and within a reasonable time of the summit, we will be able to negotiate for more time," he wrote.[40]

Working through the day and into the night, the three took turns leading the climb, pounding hundreds of bolts into the rock before retreating to the valley floor to gather more food, water, and gear. Winter came, and Feuerer and Powell halted their campaign, each exhausted from the effort. Harding remained, however, recruiting Wayne Merry and George Whitmore to finish the ascent. Finally, at six in the morning on November 12, 1958—after spending the night camped out on the rock—in one last fateful push, an exhausted Harding pulled himself over El Capitan's ledge. Equally worn and filthy, Merry and Whitmore soon joined him in celebrating, in the pale morning sun, the conquest of Yosemite's last big climb. It had taken forty-seven days of climbing, over two years, by five different men to complete the historic ascent. News lauding the three men's accomplishment splashed across the front page of newspapers nationwide, and even *Life* magazine bought photographs from the group for what would have been its cover story, had the election of Pope John Paul XXIII not captured the public's imagination.[41]

But even as Harding, Whitmore, and Merry stood triumphantly on top of El Capitan, a fierce public debate over their accomplishment raged below. "The Climbers had no more right to make spike holes in El Capitan than they would have had in carving their initials on a Giant Sequoia," raged a letter writer in the *San Francisco Chronicle*. Claiming that the climb was vaguely sacrilegious, another visitor wrote in a complaint to NPS Director Conrad Wirth: "There are some things man should leave to God without showing God he is not so great after all." Mather and the Park Service took such public grievances seriously but maintained rock climbing was an appropriate form of recreation in Yosemite. Writing to the Sierra Club, the acting NPS director, E. T. Scoyen, indicated that the climb and its accompanying media frenzy had "exerted a strong negative influence on the normal use of the park by visitors and on the normal park atmosphere."[42]

Despite the media circus that had surrounded the climb, the NPS remained reluctant to further regulate climbing within Yosemite. In the days following

Harding's summit of El Capitan, Yosemite's superintendent, John Preston, confirmed rock climbing's legitimacy within the park, writing, "We feel we should recognize rock climbing as one of the many ways people can find recreation and pleasure in our National Parks." Yet he also had grave concerns over the risks climbing presented, noting, "As a rule, we have very little difficulty with well-qualified climbers; it is the inexperienced and ill-equipped persons who get into trouble and require rescue operations which sometimes lead to unfavorable publicity for the National Parks."[43]

To solve this problem Preston called for a cooperative arrangement that ensured the safety of what he predicted would become an inevitable wave of inexperienced climbers seeking to replicate the feats of those such as Harding and Robbins. While Preston and others within the Park Service looked to continue to accommodate climbers, Conrad Wirth, the Park Service's director, took a much dimmer view of climbers within the national park. Acknowledging that he had "not had too much experience with rock climbers," Wirth railed against "trick climbing," which he defined as "driving iron pitons or bolts into a mountain," and said that it should be banned from all national parks. Further confusing the issue, Superintendent Preston seemingly backed off from his earlier support, asserting that the ascent up the Nose was "nothing more or less than a 'stunt' climb which . . . cannot be considered a legitimate form of recreation."[44]

Yosemite's climbing community, including several members of the Sierra Club, angrily rejected Wirth's declaration. Harding called the proposed ban "ridiculous" and argued that pitons and bolts in no way degraded the scenic qualities of the rock.[45] Whitmore joined Harding in his criticism, writing both Wirth and Preston, "Your implication that 'tricks' and 'stunts' were used in a carnival-like spectacle again indicates a complete disregard for the truth." He continued, "If you had taken the trouble to inquire . . . you would have found that we used standard climbing techniques which have been commonly accepted and practiced in Europe and this country for years."[46]

The storm over climbing's place in Yosemite echoed a much larger reality facing the park over the question of what was and was not an acceptable recreational activity. The ascent up El Capitan erased the fear of risk and made the impossible possible. Harding had proved that the heady combination of technology and grit could conquer even the most intimidating faces within the valley. Climbing's popularity at that time was soaring across the world. Armed with ever-improving gear, climbers of all skill levels transformed scenic buttresses of rock into playgrounds. Nowhere more so than in the

Yosemite Valley. Within five years of the first ascent up El Capitan, some 2,215 climbers visited Yosemite, numbers that reflected only those who bothered to register with the Park Service. That same year, climbers established a number of new routes: the west buttress of El Capitan, the center of Half Dome, and the Misty Wall, a granite face adjacent to Yosemite Falls.[47]

Succeeding generations of climbers developed the sport even further, capturing the attention not only of their fellow climbing community but also of those standing on the valley floor watching in awe as small specks made their way up the valley's granite walls. By the start of the twenty-first century, groups of climbers queued at the base of nearly every route within the park. Ascents became more daring and dramatic. Perhaps no more so than renowned climber Alex Honnold's free solo up El Capitan in 2017. Captured in the Oscar award–winning documentary *Free Solo*, the climb captivated much of the world, much like the first ascent up the Nose sixty years before. And like the first ascent, Honnold's garnered well-deserved praise for its spectacle. Yet his accomplishment further underscored Yosemite's allure to not just sight seekers but an ever-growing array of adventurers.[48]

As in the rest of the park, such popularity has come with an environmental cost. In 2004 longtime Yosemite climber and guide Ken Yager grew disgusted with the amount of trash accumulating at the base of most climbs within Yosemite. The most obnoxious were the hundreds of piles of toilet paper and waste. Deciding to do something about the mess, he started the nonprofit organization Facelift to clean it up. Dedicated to helping keep the valley clean, the group now runs one of the largest day events within the park, and volunteers have removed more than a million pounds of soiled toilet paper, wrappers, cans, asphalt, and even rebar from the Yosemite Valley. Such efforts have only become more monumental as more and more climbers venture into the park.

This popularity has led to overcrowding on routes once deemed too difficult to climb. One fall day in 2019, rangers counted thirty-two climbers on the Nose, more than the number of pitches that make up the historic climb. Such crowding has led some to believe that a permit system needs to be put into place, a move that runs counter to climbing's culture within the valley.[49]

OUTSIDE OF THE Yosemite Valley, Badger Pass likewise enjoyed a swell in popularity as millions of Americans hit the slopes. New ski resorts at Mammoth and the Lake Tahoe region joined Yosemite's small ski area in drawing visitors from Los Angeles and the Bay Area into the Sierra. Among them was Raytheon engineer Ben Finley. Having moved to Los Angeles after college to work as an engineer, Finley quickly embraced the California dream, taking up scuba and sailing. But it was on a dare from his girlfriend that he discovered the sport for which he would become renowned. After a Thanksgiving trip to Yosemite, she stopped on her way out of the park to watch people skiing at Badger Pass. Returning home, she told Finley she wanted to go skiing. Less than enthusiastic about the idea, he struck a deal with her. If she was willing to learn to scuba, then he would take her skiing. A month later he was enrolled in ski lessons at Badger Pass. And by the end of the winter Finley was a capable skier. More importantly, he had found what would become his life's passion.[50]

Finley spent the next two years earning a graduate degree at New York University while continuing to ski. On weekends he joined other African Americans on ski trips to upstate New York ski areas. After earning his master's degree, Finley returned to Raytheon, but, unlike in New York, he found few African Americans in Southern California interested in skiing. To address this, he organized the 4 Seasons West Club of Los Angeles in honor of the Harlem ski club with which he had skied in New York. From his home in Southern California, he began leading trips, taking fellow African Americans skiing in Yosemite and other Western resorts.[51]

Even though membership in the 4 Seasons West Club grew over the next several years, skiing, like tourists in the national parks, remained predominately white. In 1972 Finley was introduced to the director of Sno-Gophers Ski Club of Chicago, Art Clay. Seeing the possibility of expanding skiing's popularity within the African American community, the two men organized the first "Black Summit" for the following Thanksgiving week in Aspen, Colorado. The week was a success, attracting more than 350 Black skiers, and Finley and Clay agreed on the need for a national organization to promote African Americans' taking up the sport. And so the following year they formed the National Brotherhood of Skiers. Over the next fifty years, the organization introduced thousands of African Americans to skiing. In 2019 both Finley and Clay were inducted into the U.S. Ski and Snowboard Hall of Fame, an honor that had begun with that trip to Badger Pass all those years before.[52]

But for every Ben Finley, there were thousands of other African Americans who have never felt welcomed in Yosemite. Scholars have provided explanations as to why national parks remained predominantly white, even as California and other Western states became increasingly racially diverse. They include a widening economic divide between racial minorities and the growing white middle class, overt discrimination, and cultural boundaries. To this last point, it is telling that none of the Park Service's internal memos on Yosemite in the decades following World War II dealt with race. Even the Park Service's 1953 Visitor Survey failed to ask respondents to identify their race. And it was not until four years later, in its last year of publication, that *The Negro Motorist Green Book* even noted Yosemite National Park as a possible destination. Such realities reflect white cultural beliefs that African Americans and Hispanics simply did not belong in national parks. Too often, people of color had internalized this message and come to believe that parks like Yosemite were not for them, and so few visited. Such ideas drew from the very foundation of the national park ideal, which grew out of white middle class views of nature in the nineteenth century. Correspondingly, many white Americans, whether consciously or not, viewed minorities as outsiders within the parks. And so the combination of cultural and economic limitations left millions of Americans shut out of Yosemite.[53]

AFTER YEARS OF funding delays, construction on the Tioga Road's middle section finally began in the spring of 1957. Snowfall shut down work that fall, giving the Sierra Club time to redouble its efforts to halt the road's construction. Fresh off his victory in stopping the damming of Echo Park Canyon in Dinosaur National Monument, Brower believed similar tactics could be used again in Yosemite. The campaign proved effective enough that later that summer Director Wirth invited Brower, Bradley, Adams, and a collection of Sierra Club and Park Service officials to meet and discuss routing options for the controversial section of road. While touring the glacial dome, Wirth pointedly asked Brower why the Club had reversed direction on the road. Brower reportedly retorted that it was a different Sierra Club now.[54]

Still, many on the Sierra Club's board remained reluctant to directly confront the Park Service, fearing it would harm the Club's long-held relationship with the federal agency. Incensed, Ansel Adams resigned from the Club's board, dismayed over its failure to protect the Tenaya Lake region from the bulldozer. Such appeasement would not only destroy the region, he later wrote in the *Sierra Club Bulletin*, but more importantly would steal the magic he had experienced as a young man when he first encountered the lake's primal beauty. It was bad enough that the road would allow visitors to speed through the region at seventy-five miles an hour, a common exaggeration made by the road's detractors, but the scars left behind by blasting a road through tons of granite would forever destroy that beauty and solitude.

Particularly insulting to Adams was the construction of a borrow pit, which he argued was little more than an open wound on the landscape. In response to being told that the pit could not be seen from the road, Adams roared, "This would suggest that all areas of the national parks which cannot be 'seen from the road' should be returned to public domain—or to the Indians!" That such righteousness could be seen as insulting did not occur to Adams, who, like Brower, came to view the preservation of Yosemite's wild character from growing hordes of visitors as inviolable. But Wirth remained unswayed by such arguments, and construction continued on the Low Line route along the lake. The decision chilled the Park Service's relationship with the Sierra Club for the next several decades as the longtime allies came to blows over Yosemite.[55]

On June 23, 1961, Wirth stood under a perfect blue Sierra sky and addressed a small gathering at newly named Olmsted Point for the road's dedication. "Thousands of people—and eventually millions—will come up here and look down into the Yosemite Valley to gain one of the greatest inspirations possible to mankind," he declared. Wirth asserted that without its construction none but the tiniest fraction of visitors would have ever had the chance to enjoy the view, a thinly veiled barb at his detractors within the Sierra Club.

Wirth then went on to tell how, eleven years prior, he had received a letter from a friend detailing his terrifying experience traversing the old Tioga Road. While his friend could not give a description of the scenery along the old road, he had been able to give a good account of every stone and bump along the way. Likewise, his friend's wife had been unable to enjoy the view, for she had devoted much of the trip to prayerful contemplation of his driving efforts. Such experiences would no longer be the case, Wirth argued, inviting his friend to make the drive again. "After this summit experience, I am sure that his arrival in the valley will be one of greater grandeur and beauty than anything he experienced before," Wirth concluded. Now all could enjoy Yosemite's beauty and wonder.[56]

In many respects, the fight over the Tioga Road marked an inflection point in a broader conflict over the management of Yosemite. Believing the road's expansion represented an outdated, and flawed, vision of national parks that favored visitor enjoyment over the preservation of nature, a growing number of environmentalists, including Brower and Adams, argued for the need to set aside much of the national park as wilderness. Doing so, they contended, would protect the park by helping stave off the damage being wrought by its growing numbers of visitors. In this sense, environmentalists were rejecting the Park Service's expertise, believing the federal agency could no longer be trusted to fulfill its mandate to preserve the national parks by limiting recreational development within Yosemite. In order to ensure the Park Service protected natural landscapes for future generations, environmentalists believed, Congress needed to permanently designate a large part of Yosemite as wilderness.[57]

By 1956 the debate over the Wilderness Act had reached the floor of Congress. Seeking to build upon the success of defeating the Bureau of Reclamation's plan to build a dam in Dinosaur National Monument, environmentalists, led by the Wilderness Society's Howard Zahniser and the Sierra Club's Brower, began circulating a bill that would designate large swaths of public lands as wilderness. They asserted that the Park Service had not distinguished itself during the four-year political battle over Echo Park, particularly Director Wirth, who had seen the dam as an opportunity to build yet another national recreation area like Lake Mead. The architect of Mission 66, Wirth remained tied to the belief that the Park Service's fundamental purpose was first and foremost to provide recreational opportunities, a view that put him directly at odds with preservationists. So it was no surprise that Wirth took umbrage at the wilderness bill's targeting of the Park Service's management of parks like Yosemite.[58]

In a letter to Zahniser, after reviewing an initial draft of the bill, Wirth stated that he wholeheartedly favored wilderness protection but believed that the areas included in the bill were, "in fact, already, wilderness areas with adequate protection against future nonconforming use." By lumping national parks and national monuments in with other types of areas such as national forests, national wildlife refuges, and Indian reservations, the wilderness bill would diminish the Park Service's ability to meet its mandate. Furthermore, Wirth questioned the bill's placing the secretary of the Smithsonian Institution in charge of managing wilderness, noting it would undercut the Park Service's authority. He further asserted that the Organic Act already provided wilderness preservation within the national parks, making their designation as wilderness redundant and so unnecessary.

Each of Wirth's arguments underscored the Park Service's resistance to any change in its mandate, and thus its authority. The Wilderness Act not only weakened the agency's flexibility in managing the parks but recast the

Park Service's mandate as a paradoxical "dual mandate." For generations, the Park Service's Organic Act had proved a workable mandate due to the national parks' moderate visitation. But as the national parks' popularity had grown, the Park Service's ability to balance conservation, not preservation, with visitor use came into question. In this manner, the Wilderness Act, combined with the Park Service's 1963 publication *Wildlife Management in the National Parks*, more commonly known as the Leopold Report, and Secretary of Interior Stewart Udall's classification of national parks as natural, cultural, or recreational units, crafted a new narrative around the national parks that emphasized ecological preservation over visitor enjoyment.[59]

It would take years of fierce debate before Congress passed the Wilderness Act. President Lyndon Johnson finally signed the bill into law in 1964, three years after the Tioga Road's completion. Mission 66 ended with much fanfare just two years later. Yosemite's New Village, campgrounds, sewage and electrical facilities, and, of course, the Tioga Road, reinvigorated the park, making it more accessible and modern. But such improvements also fueled Yosemite's popularity. Visitor numbers surpassed two million people in 1967, double what it had been, just a decade earlier, when the Park Service initiated its Mission 66 program.

The solution to overcrowding in Yosemite and other national parks remained highly contentious. Utah Senator Frank Moss argued from the Senate floor in August of 1968 that "we are forgetting . . . that our national parks are for people, and that a vast majority of Americans see most of our natural wonders, and the greatest part of our magnificent scenery, from the window seats of their automobiles, or after a short hike from a visitor's use area." Noting that most visitors were far more interested in getting an unobstructed view of Old Faithful than finding a remote wilderness camp spot in Kings Canyon, Moss asked, "Are we overemphasizing wilderness and underemphasizing people?"[60]

For someone who had voted for the Wilderness Act just five years earlier, Moss's question was telling. To the senator, the promise of the national parks relied on the public's ability to enjoy their natural and historic wonders, and on Congress having set aside national parks expressly for the American people's use and enjoyment. However, as Moss repeatedly pointed out in his speech, this was becoming increasingly difficult to do. The solution to the growing number of visitors was not to keep the parks in a primitive state to be enjoyed by a mere few, but to make them more accessible by developing more roads, trails, and campgrounds. "The man with a trailer has just as much right in our national parks as the man with a pack on his back," Moss concluded. In fact, he had "more right really because he is in the vast majority."[61]

A growing number within the Park Service took umbrage at Moss's remarks, countering that parks such as Yosemite had become so accessible that they were in danger of being overrun. And such large numbers of visitors were detracting from the very thing for which the parks had been set aside. In a letter responding to Senator Moss's speech, Rocky Mountain National Park's ecologist, Bettie Willard, noted, "Today, some people say they would not think of visiting the Yosemite Valley during the tourist season of July and August when it is so lovely physically, but made less beautiful by crowding." She was not alone in her assessment.

Yosemite's first natural resource manager, Robert Barbee, contended that many of the nation's most popular parks were being loved to death, and to mitigate such volumes of people the agency must balance the public's desires with each park's ecological limits. Echoing the language in the Leopold Report, he argued that within the boundaries of many parks, including Yosemite, could be found priceless vignettes of primitive America. And thus, it should be the rule to preserve naturalness by leaving such natural parks unimpaired. This view would slowly become policy as a new generation of biologists began to take management positions within the National Park Service.[62]

Following the Leopold Report's release, Secretary Udall addressed the national parks' growing popularity by redesignating parks into three categories—natural, historic, and recreational. Natural areas, he stated, were the oldest category, reaching back to the establishment of Yellowstone. Such parks were directed to follow the Advisory Board on Wildlife Management's recommendation to reestablish indigenous plant and animal life, while limiting development to only what was deemed necessary and appropriate in order to safeguard each park's natural environment. By sorting national park units into the three categories, Udall effectively redefined the Park Service's Organic Act. How a park was categorized would ultimately determine how it was managed. Under this policy, cultural resources would take a back seat in natural parks like Yosemite.[63]

Such understandings had profound implications in how the Park Service managed and how visitors interacted with their parks. New laws, including the National Historical Preservation Act, Wild and Scenic Rivers Act, National Trail System Act, and Plan for Fire in Ecosystem Act, among others, would pull the Park Service in multiple directions. Combined with the twin challenges of limited budgets and ever-increasing numbers of visitors, such legislation guaranteed that no one would ever be satisfied over how parks were managed. Many visitors clamored for more access, better facilities, and fewer people, while others loudly protested any construction within Yosemite. It became a fight between access and preservation. All of this came to a head in 1970 when the Department of Interior rejected Yosemite's Master Plan, continuing the debate over access to the park.

FIG. 16. Skaters enjoy Camp Curry ice skating rink, 1933. Yosemite Historic Photo Collection.

FIG. 17. Sailors enjoying the view from Inspiration Point in May 1944. Yosemite Historic Photo Collection.

FIG. 18. Crowded parking lot at Badger Pass Ski Area, 1941. Yosemite National Park Archives.

FIG. 19. (*opposite top*) Congested campground in Yosemite Valley, 1952. Yosemite National Park Archives.

FIG. 20. (*opposite bottom*) Visitors pushing cars through a narrow section of Tioga Road in the 1950s. Yosemite Historic Photo Collection.

FIG. 21. (*above*) Camping at Tenaya Lake, 1956, photograph by Robert N. McIntyre. Yosemite National Park Archives.

FIG. 22. Visitors reading an interpretive sign in the Mariposa Grove, 1947. Yosemite Historic Photo Collection.

FIG. 23. Park Ranger explains Glacier Point in the 1960s. National Park Service History Collection.

ON THIS SITE PRESIDENT THEODORE
ROOSEVELT SAT BESIDE A CAMPFIRE
WITH JOHN MUIR ON MAY 17, 1903,
AND TALKED FOREST GOOD. MUIR URGED THE
PRESIDENT TO WORK FOR PRESERVATION
OF PRICELESS REMNANTS OF AMERICA'S
WILDERNESS. AT THIS SPOT ONE OF OUR
COUNTRY'S FOREMOST CONSERVATION-
ISTS RECEIVED GREAT INSPIRATION.

FIG. 24. (*opposite*) Roosevelt and Muir interpretive sign at Bridalveil Meadow memorializing the last conversation between the two on Yosemite's future. Yosemite National Park Archives.

FIG. 25. Illustration by Steven Johnson in Roger Olmstead's article "Why Not Rent a Park" in *Sierra Club Bulletin*, March 1973.

FIG. 26. Miwok–Kashaya Pomo demonstrator Julia Parker working with children in front of Yosemite Museum sometime in the late 1970s. Photograph courtesy of Ronald Carr.

FIG. 27. Gate into Lower River Campground on January 2, 1997, photograph by Steve Thompson. Yosemite Research Library.

FIG. 28. Hikers climbing Half Dome cables in August 1990. Photograph by John Poirmiroo. Yosemite Slide Collection. Yosemite Research Library.

FIG. 29. Ranger Shelton Johnson, dressed as a Ninth Cavalry Buffalo Soldier, addresses a congressional delegation in the Mariposa Grove. Yosemite Public Affairs Collection. Yosemite Research Library.

FIG. 30. Group of tourists enjoy view from Inspiration Point in 2015. Photograph courtesy of Leisl Carr Childers.

PART 3

5 A Flood of Visitors

The buzz of voices filled the air as people gathered in small knots scattered across the eastern end of the Yosemite Valley, waiting expectantly for darkness to come. Dozens sat in front of the stage at Camp Curry, frequently glancing up from their conversations to Glacier Point. Others had found their favorite spots in the surrounding trees, parking lots, and meadows. Softly at first, but quickly gaining volume as more campers joined in, the strains of "America the Beautiful" rose above the din. Soon it felt like everyone in the valley was singing.

The conversations and singing immediately hushed as a faint voice rang out, "Hello Glacier Point!" The crowd quieted as a single voice bellowed from above, "Hello Camp Curry!" Then another voice, typically that of the night bellhop, hidden behind a nearby boulder, yelled, "Is the fire ready?" A beat later, Glacier Point responded, "Yes, the fire is ready." The invisible voice then bellowed, "LET THE FIRE FALL!" As if by magic, a cascade of glowing embers poured off Glacier Point, making it appear as if a waterfall of fire were flowing down the granite face. As thousands of viewers gasped in delight, the sound of cameras clicking and whirring in quick succession filled the air as the vivid red coals streamed down the darkened granite face.[1]

The throngs of viewers stirred as the last embers winked out. Cars roared to life, and traffic quickly became snarled along the valley's narrow roads. Impatience set in as drivers inched their way back to their camping spots. Seeking to seize a perceived advantage, a few steered out into the grassy meadow, only furthering the chaos. Pedestrians crowded the few footpaths and spilled out onto the narrow road's shoulder. Piles of used food wrappers, bottles, and refuse littered the ground. Overwhelmed by the volume of traffic, the handful of rangers did what they could to untangle the lines of cars full of cranky visitors. Gradually the roads opened, and the Yosemite Valley fell into a fitful rest.

Wildly popular since its inception in 1872, the Firefall had grown into a beloved tradition in Yosemite. Many visitors considered it the highlight of their annual trip to the park. So when the Park Service announced in 1968 its cancellation of the event, the news came as a jolt to many. Calling the

Firefall "artificial," Park Service Director George Hartzog decided the event had grown too large, created too much traffic, and left behind too much litter. Charged with protecting the park's natural wonder, the Park Service, he asserted, could no longer in good conscience allow the event to continue. And so the event's popularity had led to its demise.[2]

Angry letters flooded Yosemite's administrative office following Hartzog's announcement, demanding the Firefall be reinstated. Many letter writers told of their earliest remembrances of the Firefall, of the evenings singing with their families while waiting in deep anticipation for the call to Glacier Point and the stark beauty of the glowing embers cascading down the darkened granite cliff. Reminiscences on those summer evenings far outweighed any concerns over the nightly traffic jams or the place of a man-made spectacle in the Yosemite Valley. Even the media piled on, with countless editorials criticizing Hartzog's decision, many arguing it would ultimately hurt the local economy, which was reliant on tourist dollars. The editor of the regional magazine *California Farmer*, Jack Pickett, wryly suggested that if the park was so concerned about attracting too many visitors it should cut down all the trees in the Yosemite Valley. Despite such blowback Hartzog remained steadfast in his verdict that the Firefall distracted from the park's true purpose, and the beloved tradition faded into memory.[3]

The Firefall's cancellation, and the ensuing backlash, highlighted the very real limitations of visitor use within Yosemite. With the park hosting more than two million visitors annually, crowds were overwhelming the park's campgrounds, roads, trails, and scenic overlooks. Rather than finding space for the quiet contemplation of Yosemite's wonder, the park's visitors more often encountered mountains of litter, traffic jams, and campgrounds chaotically packed with tents, cars, and people. Bumper-to-bumper traffic brought mounting complaints about air pollution and the lack of parking within the valley. Weekend traffic congestion had become so bad within the Yosemite Valley that Park Service officials worried that they would soon have to begin putting up "Closed to Vehicles" signs at the park's entrances. That fear became reality within the next two years, when rangers began turning cars away from the valley on busy summer weekends.

Outside of the valley, the constant stream of cars passing through the iconic Wawona Tree had so weakened the tree's root system that the giant sequoia could no longer support itself, causing it to fall over in the winter of 1969. The tree's collapse brought an end to a nearly ninety-year tradition of visitors driving through its tunnel by horseback, wagon, and automobile.

While saddened by the giant sequoia's death, one ranger voiced his relief, telling the *New York Times*, "I hate to say it was fortunate, but the tree was a real headache, a major traffic jam." Throughout Yosemite, sitting in traffic had become as common an experience as standing at the base of Yosemite Falls, leading to mounting calls to limit the number of cars coming into the park.[4]

While traffic remained the park's most visible problem, camping and lodging were also reaching a crisis point. As early as 1965 Director Hartzog described Yosemite as being a "great metropolitan area in the summertime." Overcrowding had become such a problem that by the late 1960s the Yosemite Valley had earned the unwanted, but warranted, nickname "Yosemite City."[5]

Writing in 1974 famed photographer and Yosemite climbing legend Galen Rowell noted, "Already the flood of people has damaged the wilderness character of far more terrain than has the quiet reservoir in Hetch Hetchy Valley." A reality, he argued, that should shock all out of the belief that they could continue to experience Yosemite as they had in the past. Rowell was not alone in his assessment. Many were coming to the same conclusion: that the growing crush of visitors threatened to overwhelm Yosemite. But even as park visitors agreed that Yosemite had become too crowded, most resented any restrictions on their continued enjoyment of the park.[6]

As the crowds continued to grow over the coming decade, so did the controversy over how to preserve Yosemite's natural beauty for all to enjoy. Not since the bruising political fight over Hetch Hetchy had Yosemite been so embroiled in such a heated debate over its future. Seemingly everyone had an opinion, but few agreed on any one solution. This became particularly clear when, under a tremendous amount of pressure from environmental groups, Park Service Director Ronald Walker rejected both the park's General Management Plan and Wilderness Plan in 1974, calling both "deficient in several respects." This was an enormous victory for environmental groups opposed to any further development within the park. Walker's decision thrust the problem of visitation in the park into the national limelight for the rest of the decade.[7]

WHILE THE PARK Service's cancellation of the Firefall exasperated many park visitors, frustrations with crowding in Yosemite took a much darker turn in the summer of 1970. Seeking a bit of solace from modern life by camping along the Merced River, peering out over Glacier Point, or standing at the base of the Yosemite Falls, visitors instead encountered traffic, chaos, and loud noise in the campgrounds, and both the Yosemite and Curry Villages teemed with people, including a growing number of young people. To many, the long-haired and strangely dressed youth were simply an oddity. But to others, including many longtime park rangers, the growing numbers of "hippies" roaming Yosemite Valley were an unwelcome intrusion. Park rangers fielded endless complaints about loud music, marijuana smoke, loose dogs, public nudity, and theft. "It seemed that every group had loud stereo systems, and we kept moving from one campsite to another trying to quiet them down and hoping it would stay that way," law enforcement ranger James O'Toole recalled on the rangers' struggles in maintaining order within the valley.[8]

Similarly, Don Hummel, president of the Yosemite Park & Curry Company, which operated lodging and food concessions in Yosemite Valley, grappled with the rampant panhandling, loitering, and shoplifting within the valley by hiring private security. Some tourists met the new security guards with a mixture of resignation and outrage at the presence of a private police force within a national park. Most, however, aimed their ire at the few panhandlers sitting outside the village's restaurants and shops.[9]

But if there was a single issue that bedeviled visitors and rangers alike, it was the lack of parking. Unable to find parking in campgrounds, which limited visitors to one vehicle per campsite, or in nearby parking lots, many drivers had taken to simply pulling off onto the side of the road. This, in turn, further congested the Yosemite's heavily used thoroughfares. Congestion was particularly bad near Stoneman Meadow, where mostly college-aged visitors had begun to congregate in increasing numbers to drink, smoke, play Frisbee, and hang out. Park rangers managed both the illegally parked cars and the growing crowds within the meadow by citing vehicles parked on the meadow's edge, which was often enough to encourage their owners to move on.

This informal arrangement ended abruptly on Memorial Day weekend 1970. That Saturday evening, rangers announced over megaphones to those in the meadow must remove their cars parked alongside the road or risk being ticketed. When few showed any interest in complying, rangers closed the four-way intersection at Camp Curry to stop any further traffic then called in a tow truck to begin removing the illegally parked vehicles. With the parking issue seemingly settled, Yosemite's chief law ranger, Dave Patterson, ordered rangers to begin pushing people out of the meadow, while he and three other rangers ensured none slipped back. Few in the meadow initially noticed the rangers' entry. But after the noisy arrest of a young man for pulling a knife, "the crowd seemed to explode," Patterson later wrote. Fearing further violence, he ordered his men to withdraw from the meadow. Emboldened by the rangers' departure, and seething over their constant harassment, many lingered within Stoneman Meadow before the crowd finally dissipated.[10]

While rangers had made only a single arrest, tensions within the Yosemite Valley remained high the following morning. Making matters worse, crowds continued to pour into the area, overrunning the already crammed campgrounds and stretching the nine park rangers on duty to the breaking point. "By Saturday the men were extremely tired, having only 7 or 8 hours' sleep in a 48-hour period," Patterson recalled. In addition to having contended with the crowds gathered in Stoneman Meadow, the exhausted rangers had to deal with multiple car accidents, rerouting incoming traffic out of the valley to alleviate overcrowding, and conducting what few foot patrols they could manage. While no further violence occurred, the size of the holiday crowds that had descended upon the Yosemite Valley raised alarms within the Park Service over the coming Fourth of July weekend.[11]

Such concerns proved prophetic as thousands of visitors swarmed into Yosemite for the long holiday weekend. Seeking to prevent another confrontation in Stoneman Meadow, Superintendent Robert Arnberger ordered quiet hours within the valley to be expanded to between 10:00 p.m. and 7:00 a.m. Once again groups of young revelers gathered in the meadow. By early evening an estimated three hundred had settled in, and they showed no sign of obeying the curfew. The situation in many of the valley's campgrounds was no better, with loud drunken parties shattering the evening peace. Seeking to regain control by first removing the increasingly loud parties from Stoneman Meadow, rangers began to slowly encircle the meadow. As quiet hours began, a ranger ordered the crowd to disperse over a loudspeaker while law enforcement rangers, reinforced by thirteen wranglers and packers on horseback, entered the meadow in a long skirmish line.

Walking past Stoneman Meadow with his eldest daughter that evening, John Fisher watched as the line of rangers and mounted park employees moved into the meadow. "Before my very eyes we watched these children stampeded, several being clubbed, and two thrown to the ground, handcuffed, and led off to jail," the physician and former Florida state senator later wrote in a scathing open letter to President Richard Nixon. In the ensuing chaos, Fisher was separated from his daughter, only to be reunited with her when a ranger physically dragged her to him and demanded to know if she was, in fact, his child. Incensed as much at her treatment as at the Park Service's heavy-handed tactics, Fisher insisted on speaking to the superintendent but was told that Arnberger would not be available until Monday. Seething, Fisher and his daughter returned to their campsite for the night.[12]

After engaging in a "large confrontation" with those remaining in the meadow, rangers succeeded in pushing the crowd out of the Stoneman Meadow and into Camp 14. There, they spoke with several of those who had been in the meadow about park regulations and the environmental impacts of such large crowds in the valley. A handful pressed the rangers on the rationale of removing people from the meadow to keep them from trampling grass but allowing pollution-spewing cars and motor homes into the valley, but most simply drifted back to their campsites for the evening.[13]

The Fourth of July morning began in typical Yosemite magnificence. Campers awoke to the sound of stellar jays squawking as the first rays of light danced along the Yosemite Valley's iconic granite walls. And slowly the smell of wood smoke and coffee drifted across the valley. In Camp 14 the Fisher family prepared for a hike. Still fuming at his daughter's treatment by park rangers the evening before, Fisher again sought to speak with Arnberger. But once again he was turned away, told that the superintendent would not be available until the following morning. Meanwhile, several other families packed their cars and left the park, disgusted by the previous evening's events, and fearful of events escalating. Returning from a hike later that afternoon, the Fishers once again found the meadow teeming with people. But a somber mood had settled over the area. And as the evening shadows were cast across the valley floor, rangers again ordered those in the meadow to disperse.

Standing on the edge of the meadow with hundreds of other spectators, Fisher watched in horror as park rangers, wranglers, maintenance workers, and even naturalists, wearing construction helmets and armed with axe handles, emerged again from the far tree line into the meadow. "Without any warning, the horsemen suddenly burst forth in a pack, riding the iron-shod steeds directly into the midst of the seated assembly, at full gallop, scattering all those fortunate enough not to be run over," wrote Fisher of the sight. A CBS film crew captured what happened next. Seeing horses charging at them, several within the meadow began hurling whatever lay at hand. One horse was struck on the face, causing it to panic. People scattered in every direction, and fear quickly turned into anger. Small scuffles broke out across the meadow as rangers used batons and axe handles to ward off attackers. Having lost control of the crowd, rangers beat a hasty retreat from the meadow.

But the fuse had been lit, and the valley erupted in bedlam for the next several hours. Rioters set a large bonfire in the center of the road, blocked and then rolled a police vehicle onto its side. Others skirmished with Park Service personnel for the next several hours. Fisher laid the blame for the riot directly on the shoulders of Park Service law enforcement, whose tactics he argued had incited the violence—a view that not everyone agreed with.[14]

Having arrived in the Yosemite Valley that afternoon, the Ford family had grabbed one of the abandoned campsites in Camp 14, grateful for their luck in finding such a site during the holiday weekend. Such feelings soon turned to disgust as they watched "several of the children" come "crawling out of the meadow" too drunk to walk. Parties raged throughout the campground, with jugs of wine being passed around and motorcycles racing up and down the valley's narrow roadways. Upon returning from Camp Curry, the family's three boys reported that a ranger had been attacked, his car destroyed, and that hundreds of hippies, some armed with knives, were congregating in the meadow hoping to provoke law enforcement. That evening, after attending a Naturalist Talk, the family returned to their trailer, where the sounds of fireworks, motorcycles, and screams forced them to close all of their windows in the hopes of drowning out the noise. Later, as police began to comb through the camp making arrests, the Fords awoke to screams of obscenity being thrown at the officers.

The experience remained vivid for the family months later, when, after reading Fisher's account in the newspaper, Mrs. Ford dashed off a letter to Park Service Director Hartzog blaming hedonistic youth for the riot, condemning their despoilment of the Yosemite Valley, and voicing her shock at Fisher's account of the weekend. "Never have I read such a flagrant untrue account of an event," she wrote, lambasting Fisher, for his naivete, and the failure of younger visitors to behave acceptably within the park. "We believe that the National Parks of this country are provided for the beauty and enjoyment of all," she wrote. "If the young people of this country want to enjoy the benefits of the beautiful tributes to God and a great country, they should be willing to accept the rules of decent conduct."[15]

In the end, the Park Service had called on the help of nearly 150 police officers from the nearby communities of Madera, Merced, and Fresno, along with U.S. Marshals, to quell the riot. Over the next two days, 174 people, including 41 minors, were arrested. Most of the arrests were for drugs and alcohol, but a fair number were for far more serious crimes, including assaulting a federal officer, carrying a loaded firearm, and assault with a vehicle. Five rioters were reported as needing medical care for injuries they received while resisting arrest, and another for an unspecified foot injury.[16]

More importantly, the Stoneman Meadow Riot fundamentally changed how the Park Service understood the growing national fear of lawlessness and the counterculture. For visitors such as the Fords, long-haired youth represented a clear danger to the park—a view many within the Park Service agreed with. In explaining the park administrators' actions, after the riot, Yosemite's assistant superintendent, Russell Olsen, noted that he had become used to college-age youth and their antics. "Today's fad is social protest," he noted, which had no place in a national park. To alleviate public fears of further trouble within Yosemite, Russell, with the full support of the Park Service regional office, ordered rangers to stop any visitor "under 30 with long hair" from entering the park. Rangers were to inspect the vehicles for mechanical defects and use any excuse to turn away visitors they deemed undesirable.[17]

University of California student Dorothy Goldeen and her roommate arrived at the Big Oak Flats entrance gate around midday in September of 1970, after spending five hours driving from San Francisco. While the riot had been weeks before, the park remained on edge. After glancing into the car, the ranger manning the entrance station asked Goldeen to pull to the side of the road for a vehicle inspection. Annoyed at the delay, she curtly asked why the inspection was needed. The ranger tersely explained that the inspection was necessary to prevent any accidents in the park. Exasperated, Goldeen pulled to the side, where two armed rangers begin walking around her car inspecting its condition. After briefly glancing over the car, one of the two men informed Goldeen that the light over the license plate was out, the plate needed to be bolted down, and the left rear reflector needed to be replaced before she and her passenger would be allowed into the park.

Fuming at the obvious ploy to bar them from entering Yosemite, Goldeen headed to find the nearest repair shop. After spending the next three hours getting the necessary repairs, the two women returned to the entrance station. "The same ranger, who remembered us, checked over our car," Goldeen later complained in a letter to Joseph Rumburg, director of the national parks' Western region. On seeing that the rear light was broken, the ranger once again denied the women admittance. In a fit of rage, Dorothy broke out a roll of tape and completely covered the offensive rear taillight. The light fixed, the ranger then asked to hear the car's horn in a "final, futile attempt" to deny them entrance. The horn worked, and the ranger finally allowed the two women to pass.[18]

Believing that the riots had "made it clear that the traditional methods for the administration of criminal justice have failed," Superintendent Arnberger set about reforming how the park dealt with offenders. The first order of business was to set in place a series of policies in preparation for the Labor Day weekend, in which Park Service officials feared the *Berkeley Tribe*'s call for a "10,000 freak army" to descend on the park would draw large crowds bent on protest and violence.[19]

Among the new policies was turning away all vehicles deemed "to be in an unsafe condition, or operated contrary to law" from entering the park. Evoking safety as a reason for the policy, officials told entrance rangers to find any reason to keep any visitor "under 30 with long hair" from coming in. Goldeen and others who had suffered from overzealous entrance rangers using trumped-up mechanical problems as excuses to turn them away asserted such policies were discriminatory, and that the Park Service had no authority to restrict visitors from the park based upon their appearance.

The Park Service denied such discrimination. Responding to Goldeen's complaint on her treatment by the ranger Rumburg wrote, "You may be interested to know that of the very few complaints against the program we have received, those who share your opinion that the program was discretionary universally identify with the 'under 30 with long hair' segment of our visitors." He concluded that, since she clearly would reject any justification for the inspection program, there was little else he could say other than to extend a sincere hope her trip to Yosemite had been "enjoyable, inspirational, and safe."[20]

Even as the Park Service was seeking means to stem the tide of "hippies" from entering Yosemite, many within the agency began to realize the negative implications of law enforcement's more visible presence throughout the park. An Interior Department investigation of the riot determined the confrontation between visitors and park rangers had been both unnecessary and avoidable. The report's authors cited that the crowd in Stoneman Meadow that day had shown no indication of being violent, and that park personnel had failed to follow procedure in handling the situation.[21]

The riot had also created an image problem for the Park Service. "People see the park service uniform and respond to its symbolic meaning almost automatically," Rumburg later wrote on how the riot had turned the Park Service's carefully crafted image as the protector of the nation's natural treasures into one of an armed police force. But while the addition of a sidearm, handcuffs, and a helmet presented an image "not needed for normal park functions," the rise in crime across the national park system increasingly obligated rangers to act more as police officers than as naturalists. The Park Service's new approach comforted visitors concerned about their safety, while distressing those nostalgic for the image of the friendly ranger armed with the iconic flat-brimmed Stetson hat and quick smile rather than a stern look and a holstered pistol.[22]

The agency's first step in changing its approach to law enforcement was to improve how it trained its rangers. The riots had caught the attention of Congress, which, in the wake of the 1968 riots across the country, following the assassination of Martin Luther King Jr., and the Democratic National Convention in Chicago, and now the Stoneman Meadow incidents two years later, was eager to appear tough on crime. Consequently, Park Service Director George B. Hartzog had little trouble getting funds to provide "comprehensive" law-enforcement training through the Park Service's new Federal Law Enforcement Training Center in 1971. The first of its kind, the new center institutionalized law enforcement throughout the system by creating a small cadre of specially trained law enforcement rangers to police every National Park Service unit.[23]

Believing that the National Park Service's traditional "Smokey Bear" image and "soft sell" law enforcement policy were no longer effective, Yosemite's assistant superintendent, James Olsen, ordered rangers to "tighten up their enforcement attitude." Despairingly labeled the "Yosemite Mafia" by a Park Service official in the Alaska regional office, Yosemite's law enforcement rangers embraced their reputation as an elite, if not overtly zealous, law enforcement unit. Seeing themselves as law enforcement professionals rather than ranger naturalists, the Mafia took on a zero-tolerance policy toward infractions large and small in efforts to curb the surge of crime within the park. But as the Park Service took a harder line on law enforcement by handing out citations rather than friendly warnings, visitors and park employees lamented the passing of "the old, gentle rangers," being replaced "by SWAT teams."[24]

While law enforcement remained a contentious issue within Yosemite, overcrowding continued to embroil the park in a much larger controversy over the preservation of its natural features. Despite the Park Service's best efforts, including charging entrance fees and limiting the number of vehicles per campsite, Yosemite's campgrounds, particularly those in the Yosemite Valley, remained overrun. Likewise, traffic continued to frustrate park officials as well as visitors, leading to several proposals to exclude automobiles from the valley and the construction of a gondola from Happy Isles to Glacier Point. While approving the removal of cars and buses, environmentalists attacked any proposal that added further development within the valley.[25]

AMERICANS' MOUNTING CONCERNS over the environment in the 1960s had grown into a major social movement by the early 1970s. The passage of several environmental laws, including the National Environmental Policy Act, or NEPA, which required an environmental impact statement be conducted for every federal project, cast the 1970s as the "environmental decade." In Yosemite, environmental concerns over pollution, wildlife, and overcrowding swiftly came to the forefront of public debate with the release, and subsequent rejection, of the 1974 General Management Plan.[26]

The plan sought to alleviate the problem of ever-increasing visitor numbers by balancing the enjoyment of the park and its resources with "the perpetuation of the natural processes" that had "a dynamic influence on the development of the park's ecosystems." Among the plan's initial proposals were the halting of all development within the park to its current levels and the eventual removal of all automobiles from the valley. In essence, the draft plan sought a compromise between two wildly different approaches to solving the problem of overcrowding within Yosemite. One, promoted by environmental groups such as the Sierra Club, was to remove most, if not all, development from the park. The other, promoted by the park's new concessioner, the Music Corporation of America (MCA), pushed for the addition of more visitor amenities and rejected any efforts to reduce the number of private automobiles in the park.[27]

The purchase of controlling interest of the YP&CC by the media giant MCA in 1973 reignited the debate over the role of concessions within the national park. The once family-owned concession had fallen into financial difficulties during the mid-1960s, leading to its purchase by concession giant U.S. Natural Resources, which already operated concessions in Mount Lassen, Mount McKinley, and Glacier National Parks. Yet, even under new ownership, Yosemite's concessions continued to hemorrhage cash, which its president, Don Hummel, blamed on the Park Service's resistance to allowing any investment in infrastructure within the park. Seeing an opportunity to expand its recreational division, MCA purchased a majority of U.S. Natural Resources' stock, becoming the largest concession in the national park system.[28]

MCA's acquisition of the YP&CC met with a rocky start. The corporation faced immediate criticism from multiple fronts on its operation of concessions within the park. Most damning was its role in crafting Yosemite's newest master plan. Critics accused the corporation of pressuring the Park Service to support further development within Yosemite. Many questioned MCA's calls to revive the aerial tramway from Happy Isle to Glacier Point. The proposed tramway would carry some 1,030 passengers an hour between the valley's floor and rim in efforts to alleviate crowding and, in part, to help offset lost income from the loss of the Glacier Point Hotel and Mountain House to fire in 1969. Congressmen Henry Reuss of Wisconsin and John Dingell of Colorado expressed concern over MCA's apparent influence over the Park Service, seeking "to turn this 84-year-old national park into a year-round luxury resort."[29]

Even the *Washington Post* joined in criticizing MCA, blasting the park's new concessioner for turning Yosemite into "an outdoor Disneyland," an attack both the Park Service and MCA vehemently denied. But it was difficult to argue that Yosemite had not already become an amusement park before MCA's involvement. With three restaurants, two service stations with a total of fifteen gas pumps, two cafeterias, two grocery stores, seven souvenir shops, a delicatessen, a bank, a skating rink, three swimming pools, a golf course, two tennis courts, kennels, a barbershop, a beauty shop, Badger Pass Ski Area, the Yosemite Lodge, the Wawona Hotel, and the Ahwahnee, the Yosemite Valley resembled a resort more than a nature preserve. Adding the numerous campgrounds, roads, trails, and scenic overlooks, it was clear to even the most casual observer that the park was starting to feel more like a suburb than the rustic park so many expected when they arrived.[30]

In truth, visitors had come to expect such amenities within the park, believing them crucial to their experience. Yet the blatant commercialization of the park by its concessioner did often leave a bitter taste in their mouths. Yosemite was no longer run by the family-friendly Curry Company they remembered, but instead by a corporate interloper whose primary interest was increasing its profits. It was certainly true that MCA saw recreational tourism as a growth sector, as evidenced by its subsidiary Walt Disney Productions' planned Mineral King Valley ski resort, bordering nearby Sequoia National Park.[31]

But perhaps the most egregious example of MCA's willingness to exploit Yosemite was its filming of the 1974 television show *Sierra* in the park. A spinoff of the successful crime drama *Emergency!*, the short-lived drama followed the lives of two rangers as they grappled with hapless campers, lost hikers, and a belligerent, yet endearing, brown bear named Cruncher. Lasting only three months before being cancelled, the show came to represent the National Park Service's and MCA's growing tone deafness to public opinion.

Of particular offense to environmentalists was the show's crew painting several rocks for a scene's backdrop. Environmentalists denounced the action as vandalism, befuddling the Universal filming crew, who felt they were taking every precaution to protect the park's ecosystems. "We really don't know why all these ecology freaks are picking on us," the show's assistant director complained to Galen Rowell. "They trample more meadows than we do and scar more rocks than we did by painting and cleaning some rocks to follow a script." Writing for the Sierra Club's *Bulletin*, Rowell noted that even the slowest viewer would recognize Yosemite as *Sierra's* setting, which in turn would attract greater numbers of visitors to the park. "Already subject to over use, Yosemite cannot accommodate more visitors," he concluded.[32]

WHILE CRITICISM OVER the Yosemite Valley's development dominated the headlines, visitors grew profoundly uncomfortable with the park's changing policies over the rather thorny issue of limiting their numbers by imposing camping and entrance fees. Such fees affronted many visitors used to unhindered access to Yosemite's campgrounds and entrance gates, ill feelings only further aggravated by the Park Service's and its concessioner's admitted failures in implementing an effective reservation system.

In 1974 Congress announced its intentions to end the Park Service and National Forest Service's control of campground reservations and fee collections as a cost saving measure, instead partnering with American Express and the ticket agency Ticketron to operate an automated system in which visitors use a toll-free number to secure their campsites. Turning reservations over to a private interest proved disastrous, with thousands of lost or incorrect reservations. Such was the case with the Mencini family.[33]

After calling two different phone numbers, Carl Mencini had finally been able to secure reservations for what he believed was to be a week for his family at the Lower River Campground in the valley. However, upon receiving the reservation in the mail, Mencini discovered that the Iowa-based company he had made reservations through had entered the Upper Pine campground, which was not equipped for his motor home. Exasperated, he cancelled the reservation and rebooked it for the following week at the Lower River Campground.

Believing the issue resolved, the San Jose family arrived in Yosemite to find their campsite double booked due to a computer glitch. Sympathetic to the family's plight, rangers spent the next several hours speaking with the concessioner's main office, finally swinging a deal in which the Mencinis could pay for whatever spot was available in the campground throughout the length of their stay. Over the next week, the family was shuffled from one campsite to another, under constant threat of being evicted from the park by the concessioner. Further salting the wound, once the family had finally gotten a campsite for their last two nights, they discovered the two adjacent sites were empty. While deeply appreciative of the ranger's help, Mencini fumed at his treatment. In his written complaint, he raged, "Campers should not have to spend several hours each day of their vacation moving from spot to spot. Once a camper is camped, he should not have to move if there was available space for anyone coming in with reservations."[34]

The Mencinis's experience was far from unique. Visitor complaints poured into the superintendent's office over lost reservations, camp hosts' apparent disinterest in assisting visitors, and American Express and Ticketron's failure to refund their cancelled reservations. In a biting letter to the National Park Service's regional director, one enraged visitor wrote, "Our family has been camping at Yosemite for some 25 years now, never missing a season, if for only a day or so. It has become a sort of tradition with us." Linking the reservation fiasco with other changes in the park, including the ending of the Firefall and the "heavy educational thrust by the Park Rangers," the writer bemoaned the loss of what had made Yosemite so special to his family for the past quarter century.[35]

American Express and Ticketron's inability to answer the overwhelming number of phone calls, along with experiences like those of the Mencinis, led both companies to withdraw from the pilot reservation program. Yet visitor complaints over what they believed to be the Park Service's sudden and unfair policy changes, especially camping and entrance fees, continued.

Desiderata Community School representative Marc Janowitz protested being charged thirty-six dollars for a group campsite after having received a waiver in the past as a nonprofit, tax exempt, educational organization. "Had we been informed of this policy upon our entrance we in all likelihood would have terminated our visit immediately as we have not the funds within our budget allocated for those extra charges."[36]

Acting Chief Ranger James Brady apologized but explained to Janowitz that Congress had mandated that no one using Yosemite's campgrounds was exempt from paying. Brady's explanation no doubt infuriated Janowitz, who rightly felt the policy was the Park Service's attempt to restrict the number of visitors in the valley. Such resentment continued to grow as Yosemite struggled in meeting the expectations of growing numbers of visitors, its concessioner, and Washington.[37]

The Park Service made a second attempt to create a reservation system the following summer. The new program appeared straightforward enough. To reserve a campsite, campers simply dialed a toll-free number and made the reservation over the phone. Or drop-ins could visit a campground kiosk that was to be staffed sixteen hours a day. Yet once again logistical issues plagued the new system. Visitors struggled in navigating the process, often complaining about phone lines continuously being down and the burden of having to make a reservation rather than simply being assigned a spot upon arrival, as they were accustomed to doing. Reservations were often lost, sites were often double booked, and rangers' attempts to solve such problems met with stiff resistance from concession staff unwilling to accommodate lost reservations.[38]

The public's frustrations with the continued erosion of conditions within Yosemite reached a boiling point with the Park Service's new draft General Master Plan in 1974. Thousands wrote letters decrying the proposed plan following the *Washington Post's* article comparing the park to Disneyland. Most writers railed against MCA's apparent influence in promoting further development. "I have enjoyed visiting Yosemite for many years since I was a small child, and I believe that the heavy construction that the MCA corporation plans for the Park will be disastrous to the beauty and serenity of this already too populated Sierra Area," Josephine Johnson wrote in protest of the growing commercialism within the park. No one, wrote another visitor, needed "a hotel at Glacier Point with a tramway to it, winter sports at Tuolumne Meadows, or the promotion of the Yosemite Valley as a convention center."[39]

In truth, the plan failed to endorse any further development within Yosemite. Rather, it sought to straddle the very fine line between growing demands for access and expectations that the park might remain unattended much as it always had. This balance proved elusive, however, as frustrations over limited space, crowded trails, and a lack of services continued to mount. So, eleven days before Christmas, National Park Service Director Ronald Walker announced the rejection of both Yosemite's draft management plan and wilderness plan. Calling the management plan "somewhat ambiguous" in addressing the park's stated objectives, Walker ordered a new environmental impact statement to be conducted to produce a more acceptable draft. The decision threw Yosemite into chaos for the next six years as the Park Service struggled with the impossible task of drafting a plan acceptable to all.[40]

THE RUSH OF visitors into Yosemite did more than crowd its campgrounds, clog its roads, and trample its meadows. It brought profound changes to the park's wildlife, particularly its black bear population. Simultaneously beloved and despised by visitors, black bears occupied a precarious position in Yosemite's ecosystem. For many visitors, spotting one of these gentle foragers was a highlight of their trip, a moment to marvel at nature's wildness. Yet the same animals that enchanted tourists were also vilified for their propensity to destroy tents, food caches, and even automobiles.

In 1969, prompted by the findings of the Leopold Report, the Park Service had shut down its open garbage dump. Once a popular attraction for visitors, the dump had become a symbol of the harmful effects of feeding wildlife. The closure of the dump was a necessary first step, but it inadvertently created a new issue. With their easy source of food removed, the bears turned to the valley's campgrounds and parking lots, where they found an abundance of uncovered trash cans and eager tourists willing to toss them food. The scene became all too familiar—bears scavenging for scraps, visitors leaving behind leftovers, and, in one notable incident, a camper giving an entire roast to a wandering bear. This shift in the bears' foraging habits led to thousands of dollars in damages and underscored the complexity of managing wildlife in an increasingly crowded park.[41]

Such instances underlined the growing gulf between the public's attitudes toward bears and the often-tragic consequences of such interactions. Ghastly tales, such as the accidental killing of a cub by an inexperienced ranger who failed to check the charge in his tranquilizer gun, causing the dart to blow a large hole in the animal's side, killing it, were sadly becoming all too common. With the cub's captured mother in hysterics, rangers threw the dead animal into a truck bed. Later, after skinning the carcass for use in an Indian cultural demonstration, rangers tossed the remains off the side of a cliff far from the public's view.[42]

By 1973 the practice of disposing of "garbage bears" in such a manner had become commonplace within the park. Public furor over the practice erupted that August when the spokeswoman for the local group the Extinct Species Memorial Fund, Ursula Faassii, held a news conference decrying the killing and disposal of the animals. Telling reporters, "They (bears) have just as much right to live there as people do to visit," Faassii called for the end of such Park Service policies. She was not alone. For years Park Service biologists had expressed the need to change how the Park Service managed bears in every park.

Public outcry over the bear graveyard, along with a growing understanding of bear behavior, brought about a slow shift in how the park managed its bears. However, while some visitors were sympathetic toward the plight of the park's bears, others continued to view them as a nuisance. "They destroyed my camps numerous times, and they ripped the door off my father's Mercury station wagon," explained one visitor on his support for killing all of the bears in the park. "I couldn't find parts for it, so I had to drive it with a piece of plastic taped over the hole for the last two years I owned it. I hate 'em." Accepting black bears in Yosemite, it seemed, had a ways to go.[43]

Faced with mounting criticism from both the public and park staff, the Park Service took decisive action in 1975 to address the escalating conflicts between Yosemite's bears and its human visitors. The solution was not merely to contain the bears but to educate visitors who shared their habitat. The Park Service launched a robust educational campaign aimed at teaching visitors how to safeguard their campsites from the bears and to discourage direct encounters. This initiative was especially crucial in the Yosemite Valley, where the majority of human-bear conflicts unfolded.

In addition to the park's educational efforts, practical measures, such as bear-proof trash cans, were introduced. The impact of these changes were immediately notable. Reported bear incidents dropped by nearly half, from 975 in 1975 to 516 just two years later. Injuries also saw a dramatic decrease, falling from twenty-eight to six in the same period. Despite such improvements, pressure on Yosemite's bears did not abate. Yet there was a growing recognition among visitors that they were, in fact, entering bear country. Most embraced the idea of coexisting with these animals, but the challenge of sharing the park extended beyond the realm of human-bear interactions. In the crowded spaces of Yosemite, tensions among visitors themselves led to increasing conflicts over use.

THE SILENCE OF the winter morning was suddenly broken by the cacophony of two-stroke engines roaring to life. The stench of burning exhaust quickly filled the air, as the small band of snowmobilers milled around their machines preparing to ride from Crane Flats to Tuolumne Meadows. Incensed by the invasion of noise and smell, a small group of "bewhiskered youth" began angrily hurtling obscenities at the snowmobilers from across the parking lot. While trying to ignore the jeers, Donald Brondyke hurriedly tried to get his friends and family moving toward the trailhead. But then a lone skier strode across the small lot and, in one swift motion, stooped down and flung Brondyke's fourteen-year-old son's snowmobile over before quickly retreating across the lot. Furious at the audacity of the act, but fearing to escalate the confrontation further, Brondyke quickly righted the machine and frantically motioned for the rest of his group to head out.

But after traveling for only a mile on the snow-packed Tioga Road, the snowmobilers came upon another group of cross-country skiers. This time, the skiers blocked the road, refusing to let Brondyke's group pass. Faced with the choice of either running the skiers down or turning around, the snowmobilers chose to retreat and returned to Crane Flat. Finding no rangers at the trailhead with whom to file a complaint, they drove to the park's administration building in the valley to report the incident.

There, a ranger dutifully took down Brondyke's account but stated there was very little he could do to remedy the situation. Incensed by the ranger's apparent indifference, Brondyke cancelled the group's ten-day reservation and headed home. The final insult came at Yosemite's north entrance, where one of Brondyke's party overheard a ranger mutter to another, "Good, I don't like snowmobiles anyway." Later, writing the park to complain about the encounter, Brondyke thundered, "Certainly the experience at Yosemite Park can be only considered a disgrace and a serious black eye to the Park Service," and "How can the National Park Service accept, tolerate or explain this type of park management to anybody?"[44]

In a one-page apology, Yosemite's superintendent, Lynn Thompson, agreed that both the skiers' and the park ranger's behavior had been inexcusable, and "whatever our personal opinions about hair styles, race, snowmobiles, we must treat every Park user with the dignity and grace they deserve. We fell down in your case, and for this I am sorry."[45]

Thompson's apology underscored the increasingly tense clash over snowmobiles in Yosemite. First allowing the machines into the park in the winter of 1969–1970, the Park Service believed they would increase visitors' ability to enjoy Yosemite's winter scenery without damaging park resources. "We are optimistic that snowmobilers will follow reasonable regulations [as] fully as other park users," explained Yosemite's then-superintendent, John Good. But, unlike in Yellowstone, snowmobiling failed to gain widespread popularity in Yosemite. Both unplowed during the winter, the Glacier Point and Tioga Roads offered the only possible routes for snowmobiles within the park. And while the Park Service allowed snowmobiles access to Glacier Point, most snowmobilers preferred the longer trek from Crane Flat to Tuolumne Meadows along the Tioga Road.[46]

The more contentious issue, however, proved to be snowmobiling's larger place within Yosemite. Cross-country skiers and environmental groups saw snowmobiles as incompatible with the national park ideal. The machines' "belching roaring engines, polluting the air with thick streams of dark smoke" shattered Yosemite's winter solace, frightened wildlife, and ruined other visitors' quiet contemplation of the park's winter wonder. "I could not possibly understand how such a monstrosity could be permitted in a National Park, supposedly a place dedicated to the preservation of the environment, and a place to rest our fatigued nerves from the noise and pollution of our urban environment," complained one hiker after being accosted by a group of snowmobiles on his way to Glacier Point. "This is a major invasion of wilderness, and we all must take immediate and forceful action to have the permission for snowmobiles in wilderness parks prohibited," famed photographer and wilderness advocate Ansel Adams roared in the newsletter of the local chapter of the Sierra Club.[47]

The Park Service concurred. Citing President Nixon's executive order 11644, which clarified the National Environmental Policy Act by requiring federal land managers to allow off-road vehicles such as snowmobiles only if it could be demonstrated that doing so would not adversely affect the "natural, aesthetic, or scenic values" of federal lands, Yosemite's newly appointed superintendent, Les Arnberger, banned snowmobiles from the park in the spring of 1974.[48]

A small number of snowmobilers decried Arnberger's decision as being illegal. Among those were the members of the Fresno-based Sierra Snowmobile Club, who in January 1975 sued the Park Service for excluding snowmobiling from the park. They pointed out that only those with the stamina to ski thirty-five to forty miles could now enjoy the park's unparalleled scenery during the winter. "Certainly you must recognize that one of your responsibilities is making the park available to everyone," the club noted in a letter to Arnberger. While sympathetic to the club's case, the superintendent replied that any decision to allow snowmobiles would be regarded as a major federal action with significant environmental impacts under NEPA, and so would require an environmental impact statement. The courts ultimately threw out the lawsuit, allowing Arnberger's ban on snowmobiles to stand. But as the debate over snowmobiles' suitability within Yosemite unfolded, so did a discussion over the Park Service's management of climbing within the park.[49]

CAMP 4, ONCE a tranquil haven for climbers and free spirits, had become a symbol of chaotic excess in the Yosemite Valley by the late 1960s. Overrun by self-declared dirt bag climbers, the camp was often a raucous scene of gear, drugs, alcohol, and endless late-night parties. Relations between park rangers and the climbing community grew increasingly strained as park rangers targeted climbers in efforts to impose order within the unruly Camp 4.

A fortuitous flood in the summer of 1969 provided the ideal pretext for the Park Service's intervention in the camp. Seizing the moment by claiming the camp needed repairs following the flood, the Park Service closed the upper portion of Camp 4, the area most favored by climbers. The agency then implemented a series of changes aimed at bringing discipline to the campground. Length-of-stay limits were enforced, and designated campsites were established, complete with firepits and chained-down picnic tables. A parking lot was constructed, and a $3 per night camping fee was introduced. However, the crowning blow to Camp 4's once-unruly reputation was the renaming of the campground to "Sunnyside." The name change was not merely cosmetic but a deliberate attempt to disassociate the area from the climbing bum lifestyle that had defined it for so long.

The climbing community immediately denounced the new rules. The American Alpine Club, one of the oldest and most venerable mountaineering organizations in the country, pointed out Yosemite's unique place within the climbing world, and the need for climbers to spend weeks acclimatizing to Yosemite's granite walls before attempting the park's more challenging climbs. "This, of course, has led to direct conflict with the Park Service whose function it is to allow the maximum usage by the largest number of the public who are interested in availing themselves of the Park," wrote the club's secretary, James McCarthy. While sympathetic to the AAC's complaints, Assistant Valley District Ranger Peter Thompson noted that there was no way around the fee problem and that the club "was not in the best position to knowledgeably represent the 5000 or so long-haired and hip rock climbers" in the valley.[50]

The redevelopment of Camp 4 was not the only controversy to grip Yosemite's climbing community. By 1970 only one blank wall remained unclimbed: on El Capitan, a sheer piece of granite known as the Dawn Wall. Seeing the climb as one of the last real challenges left within the Yosemite Valley, veteran climbers Warren Harding and Dean Caldwell launched an ambitious assault on the wall on October 23. Over the next twenty-seven days, the two climbers labored up the granite face, drilling 330 bolt holes in which to anchor their ropes. After an early snowstorm blanketed the valley, rangers feared the two men were stuck and launched a rescue effort. Harding rebuffed the rescue, and the climbers' efforts soon became national news.

While Harding and Caldwell basked in the limelight, their methods came under a blistering attack. Ansel Adams christened the climb a triumph of engineering that had very "little relation with the spirit of mountaineering." Former national park Associate Director E. T. Scoyen seconded Adams's sentiments, writing in a letter to his successor, Ed Hummel, "I have always felt that sensational climbing which distracts from visitor enjoyment should be shunned by climbers themselves." Both were old arguments over the spectacle of big wall climbing within the valley. As the sport continued to grow in popularity, and climbers continued to press the margins of safety, many remained concerned over the sport's appropriateness within a national park. Of particular concern were the environmental costs of drilling bolt holes and hammering steel pitons into the cracks in the rock, a concern most of the climbing community also held. Many climbers had embraced the idea of "clean climbing," or using less invasive tools such as chocks and cantilevers. But such new technology only drove the sport's popularity to greater heights, drawing ever larger numbers of climbers to the valley.[51]

By 1969 the Park Service estimated roughly 12,000 climbers had visited Yosemite, a number that only promised to grow over the coming years. Despite this, Yosemite remained one of the last national parks with no climbing guide concession. Looking to capitalize on the sport's popularity, and the absence of a guiding service, longtime Yosemite climber and guide Jeff Foott proposed establishing a mountain guide service and mountaineering school in Yosemite. While this was not a new idea, the Park Service had rejected climber Mike Harding's proposed school five years earlier, climbing's popularity had grown substantially over the previous half decade. Several renowned climbers and Park Service rangers wrote in support of Foott's application, most praising his skill as a guide and in search and rescue. While receptive to the need for a guiding service and school, the Park Service questioned Foott's business acumen, preferring instead to work with the park's concessioner, the Yosemite Park & Curry Company.[52]

In February the YP&CC opened the Yosemite Mountaineering School and Guide Service, naming veteran climber and former Park Service ranger

Wayne Merry its director. The new school offered climbing lessons for everyone from novices to experts, guided climbs, guided climbing tours to Budd Lake and Mount Lyell, equipment rentals, and special programs for visitors on mountaineering. Many within Yosemite's climbing community ridiculed what they saw as both the sport's and the park's commercialization. Many were particularly scornful of the school's "Go Climb a Rock" T-shirts. The brightly colored shirts quickly became a popular souvenir among visitors, most of whom never ventured much farther than to the YP&CC's gift shop.[53]

Climbers' criticisms of the YP&CC's cashing in on rock climbing's popularity lacked a degree of self-reflection. Many of those involved in the sport's rise in fame had cashed in themselves. Yosemite climbing pioneers Yvon Chouinard and Royal Robbins launched highly successful outdoor clothing and gear companies, helping drive outdoor recreation's booming popularity. Writer and longtime Yosemite climber John Long's books further popularized climbing within the park. The climbing legend later co-wrote the 1993 film *Cliffhanger* based on the 1976 crash of a plane full of marijuana in Yosemite, and starring Sylvester Stallone. Other Yosemite climbers found fame and sponsorships by helping sell magazines and films. Attracted to climbing's allure, along with improved gear, millions took up the sport, many of whom headed to Yosemite to climb the park's iconic granite walls.[54]

With growing numbers of climbers venturing into the park, along with the rising number of visitors heading into Yosemite's backcountry, the need for a more established search and rescue operation within the park became readily apparent. Ranger Pete Thompson came to believe that the solution lay with the very same "long-haired and hip rock climbers" he had disparaged earlier. He proposed recruiting some of the camp's best climbers by offering them on-call wages and a free camping site in either Camp 4 or Camp 12. Thompson wrote, "We have kept under our wing a group of climbers who are capable of mounting a successful rescue from virtually any big wall in the world, and who are compatible among themselves, and with the Park Service overhead who will lead the operations." Made up of the Yosemite Valley's best climbers, Yosemite Operations Search and Rescue, or YOSAR for short, became an indispensable part of visitor protection within the park.[55]

Nowhere was this more apparent than in the spring of 1973. "By anyone's measuring stick, these past few weeks have been the grimmest in Yosemite's rock climbing history," wrote Thompson on the rash of accidents that claimed the lives of eleven climbers. One of the sadder tragedies came when a member of a group of four teenagers slipped and fell off the Sunnyside

Bench. Witnesses reported seeing the boy "twisting in midair" before hitting the talus field below. The young man died on the scene, after suffering multiple fractures and severe internal trauma from the fall. In his report on the accident, Thompson fumed against the "ridiculous death" after discovering that the group actually "had a clothesline along, but they were not even using that!" Even with the growing number of such accidents, the Park Service refrained from restricting climbing within Yosemite, choosing instead to promote the sport as an appropriate and enjoyable use of the park.[56]

While the establishment of YOSAR helped establish an uneasy truce between Camp 4's climbing community and the park's law enforcement, conditions at the camp remained a constant source of frustration for the Park Service. Other campers continuously complained about dirty and unattended camps, the misuse of service station facilities, the stealing of food in the lodge cafeteria, hassles over quiet hours, and drunken parties that seemed to last for days. All of which, reported ranger James Brady, were just a part of the Camp 4 scene.[57]

AS RELATIONS BETWEEN Yosemite's climbing community and the Park Service grew into one of tense tolerance, relations between the Park Service and the Native community within the park continued to deteriorate. At issue was who was allowed to live in the New Indian Village located next to the Sunnyside Campground. The Park Service considered the handful of remaining cabins tucked among the ponderosa pines an eyesore. In truth, many of the houses were in a sorry state, but that was due in large part to the agency's purposeful neglect in maintaining the village.

When the cabins fell into disrepair, the Park Service condemned them, evicting the residents. Making matters worse for the dwindling population, park administrators restricted residence to only those currently employed within the park, a significant problem for those working seasonal jobs. Finally, in 1969, Superintendent Arnberger ordered the eviction of the village's last Native residents and burned its few remaining buildings, including four cabins, a comfort station, and the shower house, as part of a firefighting training exercise.

The park's oppressive action garnered little public outrage. If anything, the removal of the few dilapidated buildings was welcomed by those seeking to purge the valley of further development. Too often, what little connection visitors made between the Miwok and the park were the romanticized displays of the Yosemite Valley's past in the park's small museum. Conversely,

Native peoples yearned for their own remembered past, one that informed their deep connection to the valley, a connection the Park Service had long denied. The demolition of the New Village simply followed the Park Service's long-held policy of Indian removal from all of the parks. Without federal recognition, the descendants of the Ahwahnechee remained powerless in opposing the Park Service's policies.[58]

By the late 1960s the Red Power movement had galvanized Native communities across the country in exerting their sovereignty and Native Nationalism. Protests such as the occupation of Alcatraz Island in the San Francisco Bay, fish-ins throughout the Pacific Northwest, and the publication of Dakota historian Vine Deloria Jr.'s bestseller *Custer Died for Your Sins: An American Indian Manifesto* had drawn wider attention to the treatment of Native peoples across North America, particularly their exclusion from their ancestral homelands.[59]

In Yosemite, the Native community began meeting to discuss how they could assert their ancestral ties to the park. Soon after the razing of the New Village, a group of Native activists formed the American Indian Council of Mariposa County (AICMC) to give voice to their continued mistreatment by the Park Service and seek ways to garner federal recognition of their connection to the valley. Despite their calls for greater inclusion in park management decisions, the intertribal AICMC struggled to gain traction with the Park Service until, in September 1976, the ongoing controversy over the park's General Management Plan provided the group leverage in securing a meeting with park officials to discuss a wide range of issues concerning Yosemite and the restoration of the Pioneer Cemetery, the resting place of many of Yosemite's earliest settlers.[60]

Like many critics of the plan, the AICMC sought a greater say in how the park was managed. Specifically, the council pressed Superintendent Arnberger on the long-standing complaint of the park's failure to hire Natives into supervisory positions. Arnberger responded that no qualified Native applicant had "yet surfaced," and that his hands were tied due to strict civil service guidelines. Council members countered that the Park Service had historically shown little interest in promoting the few Natives who had found work in Yosemite, and federal hiring regulations should be waived for long-term Indigenous employees. Additionally, they emphasized that the valley remained a "sacred shrine" for many California Natives and requested that park officials designate a new, more private space within the valley as a ceremonial area.[61]

A week later, Arnberger responded to the AICMC's demands by establishing a program to inform the council of all employment opportunities within the park. The following summer, he invited local Native high school students to participate in a two-week program emphasizing Indian cultural ties to Yosemite. While happy to see greater efforts by the Park Service to include Native peoples, council members continued to demand a greater say in how the park interpreted the history of the Yosemite Valley. The Park Service rebuffed the suggestions out of hand. With neither side willing to trust the other completely, relations between the park and local Indigenous peoples remained strained.[62]

Difficulties appeared to come to a head in 1978 when the AICMC submitted a scathing rebuttal to the park's latest draft management plan. Frustrated over being exploited as a tourist attraction, they pointed out that their elders had tried to use the Indian Gardens for ceremonies but found the "atmosphere was not proper for these traditional activities." Council members demanded the park redesignate the former Indian Village site as the ceremonial area. Additionally, they suggested the Park Service place a greater emphasis on Native history by allowing only Indigenous experts to interpret Native cultural artifacts, creating new interpretive exhibits, and enlarging the existing Indian Room. Finally, the council returned to the continued lack of employment opportunities and ceremonial access to the park afforded Indians and demanded that their future weigh heavily in the new General Management Plan.[63]

While increasingly sympathetic to Native cultural ties to Yosemite, Arnberger refused to relinquish any control over interpretation. Yet he was flexible on the issue of creating an Indian Cultural Center. After meeting with council members Jay Johnson and Les James, Arnberger directed planners to include a new cultural center in the new management plan. While pleased with the decision to include a center, the AICMC was less than thrilled over its placement at the base of the Three Brothers. In a tersely worded letter to Arnberger, the council's attorney, Barbara Karshmer, noted that the park's chosen site was "devoid of any special cultural significance" and reiterated the demand that the new center be built on the site of the former Indian Village. She concluded by noting that the "Indians of Yosemite had already been relocated twice by the National Park Service" and urging the park to "forgo yet another relocation."[64]

Exasperated at the council's apparent change of mind, Arnberger responded that the former Indian Village was unacceptable due to its location adjacent to the popular Sunnyside Campground, the heavily traveled Northside Drive, and the Yosemite Lodge and concessions employee housing. Furthermore, there was little more that he could do at such a late date, as the draft plan would soon be released to the public, followed by a series of public meetings from which a final revision would be written. Arnberger wrote, "We would expect the American Indian Council of Mariposa County to present their recommendations on the plan at those meetings along with all other groups who are equally concerned regarding the planning and management of Yosemite."[65]

Native access and permanent residence within Yosemite Valley remained the crux of the issue. While willing to work on granting greater access to the valley for religious use, the Park Service continued to refuse to allow long-term residence within the park. Under siege to remove development from the valley, the park was in no place to allow the construction of permanent housing. In contrast, the AICMC argued that, compared to other user groups, if such a term even applied, they had much deeper spiritual and historical attachments to the valley, which superseded all other claims. The council struggled in asserting that Native peoples represented more than just visitors, and so they should be given differential treatment. However, while the Park Service agreed that Native peoples had strong ties to places like the Yosemite Valley, they claimed that they could not treat Native communities any differently from any other user group.

Yet relations between Yosemite and the AICMC did thaw with changes in federal law, specifically the American Indian Religious Freedom Act (1978), and research done by Yosemite's ethnologist and museum director, Craig Bates, on the Yosemite Valley's religious importance to the various Native peoples who continued to reside around the park. Bates's research legitimized Native land claims, which gave the AICMC a leadership role in pressing its demands. The passage of the Archaeological Resources Protection Act of 1979 further improved relations between the Park Service and the AICMC by providing a legal manner to stop visitors and park employees from vandalizing or destroying Native cultural resources. Superintendent Arnberger praised the act as a way to preserve "an irreplaceable part of Yosemite."[66]

As the fight over Native access remained unresolved, so did the struggle over development within Yosemite. Following the collapse of its management and wilderness plans, due to the public outcry, the Park Service announced that it would start again on drafting a new plan. Sensitive to the overwhelming criticism that Yosemite had become a private resort developed and controlled by corporate interests only concerned about profits, Director Walker mandated the Park Service conduct a more thorough environmental impact statement that analyzed alternatives for both the preservation and use of the park and allow the public to participate throughout the process.[67]

THE REVISION PROCESS began the following spring with a series of public meetings all across California. Rather than present a plan, Park Service officials asked for and received suggestions on how to ensure that opportunities within Yosemite reflected the expectations of a broad spectrum of visitors. Also, rangers distributed workbooks to visitors in the hopes of better understanding who planned to visit the park. Most respondents noted they had spent at least one night camping in the park, and the overwhelming majority had driven their cars to Yosemite. Most strongly agreed on the need to reduce crowding within the park, but opinions of what constituted crowding varied widely. Nowhere was this more apparent than in the thousands of workbooks the park handed out to everyone who had an opinion about Yosemite's future.[68]

Two general views on how to best manage the park emerged from the public's comments. The first generally supported reducing the number of cars within the valley, but advocated keeping the drive-in campground, commercial shopping, and even the administrative buildings in place. The second sought far more substantial changes to the park. These included the removal of park headquarters and warehousing facilities, as well as several campgrounds and other accommodations, from the valley. Furthermore, many commenters sought to ban automobiles outright from the Yosemite Valley, allowing only shuttle buses, bicycles, and horses. Overall, there was a general agreement against banning automobiles completely from the park. Conversely, few responders wished to see more lodges constructed within the park.

Interestingly, out-of-state visitors strongly advocated for automobile restrictions and curtailments of present commercial activity far more than Californians. While both directions evoked strong opinions, as Park Service planner Ann Bowman pointed out, "The workbooks are not votes, and this is not an election. The final plan will be a combination of many elements." This reality ensured no one would be completely satisfied at the outcome.[69]

6 Absolutely Democratic

It had been nearly two decades since James Dodds had last visited Yosemite when he drove into the park in the summer of 1977. Much had changed in the park over that time, and not all of it to Dodds's liking. As he entered the Yosemite Valley, he grew increasingly alarmed over the traffic snarling the narrow two-lane road, how the idyllic campgrounds of his youth now teemed with visitors, and the crowds standing two to three deep that made viewing most of the park's most iconic scenic overlooks nigh impossible. "I shudder to think what the Valley floor will look like in 1997," he later reflected on his visit. Dodds attributed the dramatic environmental changes he witnessed to the incredible number of automobiles inundating the park. To solve this problem, he suggested building a parking lot at the park's entrance nearest to the valley, so perhaps one day it would regain the beauty he had first experienced in 1958.[1]

Dodds was not alone in voicing his opinion on how to best stop Yosemite's deterioration. "If people want to live as they do in the city, let them stay in the city," wrote another visitor, capturing the sentiments of many who remembered a more serene past in the park. Others, however, held very different opinions on excluding vehicles from the park. "Trying to cut out the use of our cars to go there [Yosemite National Park] would be utter nonsense," protested one letter writer, who, after wryly remarking that not everyone was a backpacker, rhetorically asked, "As for the spoiling of our environment, who are you saving it for?" Everyone, it seemed, had an opinion when it came to Yosemite. Visitors opposed the allowance of campfires anywhere within the park, called for more biking trails, complained about the slum-like appearance of the park's campgrounds, demanded extended hours for the cafeterias, clamored for more picnic grounds, and pleaded to bring back the Firefall.[2]

The heart of the problem was that Yosemite simply could not be everything to everyone. This reality had led to the secretary of the interior's rejection of the park's proposed management and wilderness plans in 1972. Sensitive to the criticism that Yosemite had become a resort developed and run by corporate interests only concerned about maximizing their profits, in 1974

Park Service Director Ronald Walker mandated the NPS conduct a much more thorough environmental impact statement analyzing the alternatives for the preservation and use of the park. This would be accomplished by allowing the public to voice their opinions on what Yosemite's future should be.[3]

Two very different narratives emerged as the Park Service continued to hold public meetings. One called for halting all commercial development in the park, removing automobiles from the valley, and scaling back the numbers of visitors entering the park to preserve Yosemite's ecosystems. The other asserted that Yosemite's principal purpose, and by extension that of all national parks, was to act as a pleasuring ground for all. Such pro-use advocates argued for accommodating the growing number of visitors by providing more amenities, including campgrounds, parking lots, and the now infamous tramway between Glacier Point and Happy Isles. Conversely, preservationists argued that Yosemite should be understood as a wilderness in which any commercial exploitation was inappropriate. No single issue exemplified this gulf more than the struggle over cars within the Yosemite Valley.

Citing automobiles as "the single greatest threat to the enjoyment of the scenic qualities of Yosemite," the Park Service remained committed, at least in print, to removing all private vehicles from Yosemite Valley. But doing so proved to be difficult. One possible solution was the expansion of the park's shuttle system. Introduced a decade earlier, the free shuttles were so popular that they frequently filled beyond capacity during the summer season. On weekends drivers were often forced to leave visitors stranded at bus stops. Yet the Park Service sought to build upon the shuttle's popularity by expanding the system to El Portal, Crane Flat, and Wawona. Another proposed solution to the park's traffic woes was closing the valley to all private cars and requiring visitors to park at either El Portal or the western edge of the valley, accessing the eastern valley via a free shuttle instead. The establishment of the Yosemite Area Rapid Transportation System, or YARTS, in 1992 appeared to create such a system, though budgetary issues and visitors' desire to drive into the valley foiled efforts to limit traffic into the Yosemite Valley in any meaningful way.[4]

Environmentalists initially supported the idea, seeing it as a means of removing all private vehicles from the valley. Conversely, both MCA and automobile advocates opposed any proposal that would keep visitors from driving their cars into the park. Writing in the *Automobile Club News Pictorial* magazine, club member Harry Coulter raged against "strict environmentalists" who "would like to see automobiles banned entirely from the park." To allow such groups to determine who should and should not be permitted to visit Yosemite, and what they should do once they arrived, "would be a tragedy of the first magnitude."[5]

With no clear consensus, and a general sense of exhaustion over the decade-long struggle over how to best manage the park, the Park Service finally approved a General Management Plan for Yosemite in September 1980. The plan called for removing all cars from the valley and Mariposa Grove and moving nonessential development to the park's periphery, as well as the preservation of wilderness so that "visitors can step into Yosemite and find nature uncluttered by piecemeal stumbling blocks of commercialism, machines, and fragments of suburbia." In introducing the plan, Yosemite's new superintendent, Robert Binnewies, noted that Yosemite had reached a crossroads: "Today, the Valley is congested with more than a thousand buildings—stores, homes, garages, apartments, lodging facilities, and restaurants—that are reflections of our society."[6]

While this was true, visitor enjoyment within the valley reflected more than just the nation's growing suburbanization. The number of visitors venturing into Yosemite's backcountry continued to swell. Hikers overwhelmed popular hiking trails, forcing the Park Service to limit the number of people allowed into the backcountry through a permit system. Like hiking, fishing remained popular within Yosemite. But by the mid-1970s the California Department of Fish and Game ended its program of stocking the park's ponds and streams by airplane, diminishing anglers' odds in Yosemite's remote lakes and streams. By the end of the decade, the state recommended not stocking any lakes that had no record of prior stocking or were found to contain no fish, in order to preserve native fisheries. While anglers grumbled that stocking added to their enjoyment of the park, the opinion that leaving as many lakes and streams untouched as possible was better for Yosemite's ecosystem held sway.[7]

PERHAPS NO RECREATIONAL activity created more furor within Yosemite than hang gliding. Invented in Southern California in the early 1970s, the sport's popularity spread quickly across the country. From Hawaii to the Catskills, pilots, as they were known, took to the air, soaring off mountainsides, mesas, and canyon walls for upward of thirty minutes before coming back down to Earth. By 1976 manufacturers had sold some 40,000 hang gliders nationwide, and advocates claimed that upward of 100,000 people had taken to the air in hang gliders. However, environmental groups derided the new sport, claiming it disrupted birds in flight and impaired other recreationalists' experiences, especially in national parks. The Department of the Interior appeared to agree with such assessments. Bucking over a half century of precedents, Assistant Secretary of the Interior Nathaniel Reed argued in 1975, nearly sixty years after the Organic Act, that national parks were not meant for general recreation and banned hang gliding. A year later the department reversed course and agreed to allow hang gliders to fly in Yosemite under very specific guidelines.[8]

Overseen by park ranger Rich Romero, the park allowed twelve flights per day. Each was required to launch by 8:30 in the morning and land within the hour to keep crowds of spectators from gathering below. Furthermore, flights were only allowed from June through October. Romero kept a strict eye on safety, monitoring every flight, and, as with climbing, helped create a unique relationship between the new sport and the national park, though one with a much lower profile. Budget cuts in the mid-1980s forced the park to turn oversight over to volunteers, a system that remained in place through the opening decades of the twenty-first century. To date, there have been over 12,000 flights in Yosemite with no fatalities.[9]

Upset by the Park Service's allowance of hang gliding within the Yosemite Valley, environmentalists argued the sport was incompatible with the park's preservation. "I don't understand the reasoning of the National Park Service to allow hang gliding from the top of Glacier Point," complained one California man. "I understand that a National Park is preserved for one's enjoyment of nature, but what if that right encroaches upon someone else's right to enjoy what the park was primarily set aside for . . . Where is the line drawn?" Superintendent Binnewies responded to the letter writer that hang gliding was both an accepted and heavily regulated use of the park, and that as a government official he must respect the rights of both hang gliders and those offended by their presence within the Yosemite Valley.[10]

Hang gliding's allowance in large part stemmed from a combination of ranger involvement and the hang gliding community's willingness to cooperate with the Park Service. Unlike with snowmobiles, the sport did not directly conflict with another user group and generally fit within the contemplative enjoyment of national parks. Many visitors loved the sight of a bright-colored glider silently soaring above the valley floor in the early morning light, and there is perhaps no better way to experience the Yosemite Valley's beauty than from above.[11]

In truth, hang gliding demonstrated visitor's evolving recreational uses of Yosemite. The Park Service historically had allowed a wide range of recreational uses within Yosemite. From downhill skiing to golf, and of course rock climbing, visitors had long understood Yosemite as a recreational playground. Yet shifts in Park Service policy and growing protests by environmental organizations challenged recreation's role within national parks. The 1980 management plan required all recreational use within Yosemite to be related to park features, but what was determined to be related appeared arbitrary. This meant the closure of the Ahwahnee golf course but exempted skiing at Badger Pass as a "traditional activity." Golf, tennis, and swimming at Wawona were also exempted since they were part of the "historic resort complex." That golfing at Wawona was "historic," and therefore allowed, but was deemed unnatural in the Yosemite Valley, and so banned, demonstrated the Park Service's struggle in balancing recreation demand with ecological restoration, especially when it had allowed recreational activities such as golfing and skiing for decades.[12]

Gateway communities watched the Park Service's slow restriction of recreational use within Yosemite with a growing sense of unease. Encompassing much of the western half of the park, Mariposa County had long had a difficult relationship with the Park Service. Locals often felt that the agency ignored the reality that their livelihoods relied on tourist dollars when making management decisions. With the Park Service's ongoing attempts to remove Wawona's inholders as a backdrop, the county's board of supervisors chairman, William Moffitt, charged that the park was once again blatantly ignoring the county by failing to provide them with a copy of the draft plan. "It is always comforting to know that the people directly concerned and who know what is going on are never given an opportunity to contribute some basic facts when this type of scare report is being developed," wrote the board chief, William Moffitt, to fellow board member Chairman Jack Brooks. "To each visitor the 'Yosemite Experience' is different," he wrote. "To one it might be catching a trout in the Merced River or a high mountain lake of the backcountry, to others it might be a round of golf, a horseback ride, hiking, relaxing in the Ahwahnee, visiting the pioneer village, skiing or whatever one might choose to do that is moral, legal, and not destructive to themselves or the Park—yes, even watching the 'firefall' that no longer exists."[13]

Similarly, Mono County officials voiced their concerns that the Park Service would curtail traffic on Highway 120 through the park. Any curtailment of traffic on the Tioga Road, Mono Chamber of Commerce President

Madalene Good argued, would substantively affect the economies of the entire eastern slope of the Eastern Sierra, "perhaps to such an extent that businesses would not find it possible to continue serving those tourists who would continue to come to the East entrance." To both Mono and Mariposa Counties, any reduction in tourism, especially automobile tourism, was an economic anathema. And so the National Park Service, they argued, needed to look beyond Yosemite's boundaries when making decisions that may reduce visitor numbers.[14]

Beyond fears that the Park Service's actions would reduce tourism dollars flowing into local coffers, county officials were increasingly angry over the treatment of inholders within the park. Conflicts over private property were nothing new to Yosemite; the park's origins were grounded in James Hutchings's legal battle over who owned the valley floor. Yet shifting federal policy toward public lands in the 1960s, from transfer to retention, targeted private landholders within the national parks. This shift particularly affected the small community of Wawona Village.

TUCKED IN THE ponderosa pine forest above the Wawona Hotel, the scattered collection of private cabins and homes of Section 35, or Wawona Village, appeared an odd place for the Sagebrush Rebellion to take root. Established in 1855 by Yosemite's first guardian, Galen Clark, the community had remained outside of the park's boundaries until 1932, when the NPS acquired the Wawona Valley. While the Park Service offered to purchase all of the private property in the area, most owners declined. For the next two decades the residents of Wawona Village quietly went about their lives in the park. As it was hidden from view, most park visitors unknowingly drove past the small residential community on their way to either Mariposa Grove or the Valley.

As a child, Charles Cushman spent most of his summers in Wawona. His father Dwight, a Los Angeles schoolteacher who worked as a seasonal ranger in the park, had bought a cabin in Wawona Village in 1961 so the family could spend their summers in Yosemite. The park was a magical place in the 1960s, particularly for a young boy. Charles often tagged along with his father on nature walks through the Mariposa Grove and became enamored with both the national park and with nature. He joined the newly founded Student Conservation Corps, dreaming of following in his father's footsteps in becoming a park ranger. But that dream ended when the Park Service targeted the scattered homes within the Wawona Village for removal.[15]

While the Park Service tolerated the few inholdings left within the park, like those that made up the Wawona Village, Yosemite's administrators grew increasingly frustrated over their lack of authority to regulate inholders' activities, as well as the need for providing law enforcement, fire protection, sewage and garbage disposal, and traffic control for the small community. Yosemite was not unique in its struggles with inholdings. Expansion of older parks, including Yellowstone, had created a checkerboard of private inholdings within their boundaries, while newer parks such as Redwood National Park and Point Reyes National Seashore were carved wholly out of private land. Controversies such as the logging of private land within Redwood, and the continued operation of the historic oyster farms within Point Reyes, had led the Park Service to view inholdings as "a serious and growing threat to the integrity of the National Park System." To meet this threat, the agency began targeting private residences within the park. Among the policy's first casualties was the Cushman family's idyllic retreat.[16]

In 1962 the Park Service notified Dwight Cushman that he had two options. One, sell his property to the Park Service and keep his job or, two, not sell and lose his position as a ranger naturalist. Resigned to their situation, as they didn't see a way to fight back, the family sold the cabin. Dwight eventually resigned his position as a seasonal ranger, and the family bought another inholding, in the Klamath National Forest in Northern California. After taking ownership of the Cushmans' cabin, the Park Service used it as employee housing for several years before eventually burning it in a firefighting training exercise.[17]

Over the following decade, the Park Service purchased, or condemned, many of the remaining private cabins within the Wawona Village, leaving a scant 204 acres in private hands. Seeking to recapture the magic of his childhood in Yosemite, Charles Cushman purchased a cabin within the historic village with a friend in 1970. The two primarily used the property for family vacations, renting it to visitors when they were not there. It was during one of those stays that Cushman observed the charred ruins of cabins as he drove into the neighborhood, evidence of the Parks Service's continued efforts to eliminate what few inholdings remained in the park. Still resentful over how the Park Service had forced his father to sell the family cabin years before, Cushman formed the advocacy group Citizens to Save Wawona to push back against the Park Service's efforts.[18]

Citing unwanted advances by park personnel, threatening letters, and the failure to approve construction permits promptly, the group began attending county meetings protesting what they believed to be the Park Service's attempts to intimidate them into selling their property. Furthermore, they noted the agency's apparent support of the Curry Company's proposed development within the park. The Park Service defended its actions, citing concerns over the area's development, and noted that the Curry Company's development was simply a proposal that was on hold until the completion of Yosemite's Master Plan.[19]

The already sour relationship between Wawona inholders and the Park Service collapsed in 1977 when, after touring Grand Teton National Park in Wyoming, Chairman of the House Subcommittee on National Parks and Insular Affairs Phil Burton concluded that the large number of inholdings scattered throughout the national parks had to go, an opinion shared by both the Park Service and the Sierra Club. Upon returning to Washington DC, Burton joined fellow Western congressmen in adding an amendment to the House Omnibus National Parks Act requiring the Park Service to acquire all national park inholdings within the next four years. Soon, inholders began receiving letters threatening condemnation if they did not accept the Park Service's offer to purchase their property.

The Department of the Interior further aggravated the situation by ordering park superintendents to halt all construction within inholdings throughout the national park system. Wawona resident Jim Downey quickly ran afoul of the new regulation when he sought to remodel the bathroom and add a storage room to his home. After park administrators told him that if he proceeded with the remodel his property would be condemned, Downey approached Mariposa County officials for clarification as to who held authority over building permits within Section 35. County officials responded that the Park Service held no authority over building codes within Mariposa County, rekindling a long-standing dispute over who held legal oversight over the small community. While Downey ultimately agreed to sell his property in exchange for a lease that allowed him and his wife to live in their home until their deaths, others were not as lucky. After three other inholders refused to sell, the Park Service simply condemned their properties and evicted them.[20]

"It's a constant cloud over your house and property," one resident complained on the stress and worry of dealing with the federal government. Cushman was less diplomatic about the situation, telling *Newsweek*, on the plight of inholders within the national parks, "We are the Indians of 1978. Except we're being kicked off the reservation, instead of being moved from one reservation to another." Mirroring Cushman's rhetoric, Wawona Village resident

Wanda Moore told reporters, "It's terrible. Now I know how the Indians felt." Equating their plight to that of Native Americans, particularly as Miwok residents were fighting the Park Service's efforts to burn down their homes in the Yosemite Valley, was offensive, but underlined inholders' rage. Years of animosity quickly boiled to the surface as the Park Service became more aggressive in removing all inholdings from the park. Believing themselves victims of an agency run amok, inholders joined the property rights movement sweeping across the West, as rural communities protested federal land policy. "We're not against parks," Cushman explained, "we're against the cavalier, arrogant way in which the National Park Service goes about buying land."[21]

Seeing no other way to stay the Park Service's actions, community members filed a lawsuit in federal court seeking to force the agency to halt the purchases and condemnations for the next three to five years. While the lawsuit ultimately failed, Cushman's reputation as a grassroots organizer and property rights advocate garnered him national attention. He soon joined other disgruntled inholders from across the country to form the National Parks Inholders Association. As the association's volunteer director, Cushman began traveling from park to park in his Pace Arrow motor home advocating for property owners' rights.[22]

Earning the nickname "Rent-a-Riot" for his aggressive style and rhetoric, Cushman was not alone in his frustrations with federal land managers. Across the American West, ranchers, private property advocates, and conservative politicians joined in what became known as the Sagebrush Rebellion. At its core, the movement sought to limit federal control of public lands. Believing that the federal government had overreached in its management of public lands, so-called sagebrush rebels pointed at growing environmental legislation as the primary threat to the West's economic prosperity—a view shared by the newly elected Reagan administration. Cushman's notoriety in fighting the National Park Service gained the attention of Secretary of the Interior James Watt, who, to the horror of many in the Park Service, named Cushman to the National Park System Advisory Board.[23]

Watt's appointment of Cushman to the National Park System Advisory Board reflected the growing philosophical and political schism over the management of the national parks. Parks, Watt contended, were not meant only for those fit enough, affluent enough, or adventurous enough to enjoy the rigors of the backcountry, but for the enjoyment of all. "Let's be realistic," he explained to the Outdoor Writers Association of America in 1981, "the vast majority of Americans who want to share in the scenic, wildlife and other natural treasures of the Nation are not rugged young backpackers." And so the national parks should be managed for the enjoyment of the majority of visitors, he asserted, who desired not only to revel in the park's natural beauty but also enjoy the comforts of modern society.[24]

Environmental groups stood aghast as Watt promoted use over preservation within the parks. "Watt seems to be having dreams of doing a new Mission 66 of opening up new routes of access within parks," former Friends of the Earth Director David Brower charged. To those like Brower, the national parks were meant to be managed not as parks but as wildernesses, a view shared by a growing number of the rank and file within the Park Service. The gulf between these two views reflected a much larger wrestling match over the Reagan administration's environmental policies. Decidedly pro-extraction, the administration viewed preservation efforts such as the Wilderness Act as antithetical to the purpose of public lands. Watt saw use, be it recreational or extractive, as the highest and best purpose of the nation's vast public lands. To that end, he took a hard stance against the Park Service's preservationist policies of the previous two decades, proclaiming that when facing a decision on how to manage national parks, "I will err on the side of public use versus preservation."[25]

Inholdings embodied this division. Writing in 1983 Western writer and historian Wallace Stegner attacked inholdings, in his now famous essay "The Best Idea We Ever Had," as "a gun pointed at the head of the parks that enclose them." While often quoted for its celebration of the national parks, Stegner's essay assailed what he saw as Secretary Watt's pro-privatization stance toward parks as counter to the meaning of the national parks as places of contemplation and solace. "Mr. Watt, like many of those groups whom he backs, sees the national parks as resorts," wrote Stegner. Inholdings were just one of the more egregious examples of private interests within the parks.[26]

The fight over Wawona Village remained unresolved for the next several years, with neither side willing to yield. That came to a sudden, and controversial, end in 1983 following the visit of Queen Elizabeth II and her husband, Prince Phillip, that February. Storms battered the Sierras as the couple toured the park with Superintendent Robert Binneweis. Like many visitors the royal couple stood in awe at Inspiration Point, visited with Julia Parker and her daughter Lucy in front of the park's museum, and witnessed the power of the Yosemite Falls during a storm.

Following the queen's departure from the park, Binneweis asked to meet with Cushman to discuss the possibility of purchasing the park's remaining inholdings. Binneweis had spent the past two decades working his way through the ranks from a park ranger in Yellowstone to becoming Yosemite's superintendent in 1979. It was a dream job that he hoped to hold on to as long as he could. Binneweis had no interest in seizing inholdings through eminent domain. In truth, the Wawona Village was not at the top of his list of matters he was juggling in managing the park. Yet he agreed with law enforcement ranger Marshal Connelly to secretly record his conversation with Cushman in hopes of entrapping the outspoken inholder. The decision proved to be disastrous for Binneweis.

A whistleblower report on the recording drew nationwide attention to both Binneweis and the Park Service's campaign to remove inholdings. In 1983, after a Department of the Interior investigation, the Park Service removed Binneweis as superintendent, reassigning him to a desk job in San Francisco. Binneweis retired soon afterward, and Yosemite's law enforcement suffered yet another public relations black eye. The issue of inholdings, however, remained unresolved.[27]

ANSEL ADAMS DIED on April 22, 1984. Few had understood the potential of Yosemite to transform a mere vacation into a deeply emotional experience and turn a tourist into a conservationist as had the renowned landscape photographer and environmentalist. Perhaps more than any other individual in the twentieth century, the eighty-two-year-old Adams had drawn the public to the national parks, especially Yosemite, and in the process defined their meaning. "In writing to you, I almost feel I am writing John Muir, or Yosemite Valley itself," one admirer wrote Adams on the photographer's influence on his life. The Sierra Club's associate executive director, Brock Evans, likewise professed to Adams, "You are in a most direct way responsible in large part for my love of the land and my passion for my job"—a sentiment shared by millions worldwide.[28]

In 1988, four years after Adams's death, the makers of the documentary *Yosemite: The Fate of Heaven* asked a young couple struggling to pitch their tent in an overcrowded campground why they had traveled to Yosemite. Annoyed at the crowds, the woman answered, "The truth is that I had gone up to San Francisco to see *Ansel Adams: One with Beauty*. I had really wanted to see that [she then pointed to El Capitan across the valley] with my own eyes after seeing it through his eyes." She was not alone. Yosemite continued to draw an increasing number of visitors seeking to see what Adams had captured through the lens of his camera.[29]

While those millions sought to experience the grandeur depicted in Adams's photographs, they also wanted a place to camp, park their RVs, buy groceries, and maybe a meal out. They also enjoyed familiar resort amenities such as swimming pools, golf courses, and, yes, even an art gallery. The combination of such amenities with the park's beauty remained a potent mix in drawing ever larger numbers to Yosemite. Yet, by the late 1980s, federal budget cuts, combined with continued growth in visitation, had left much of the park in disrepair. Poorly designed restrooms and a now twenty-six-year-old temporary concession building had rendered Glacier Point barren. Similarly, the scattered rocks at the base of Lower Yosemite Falls were often inundated with visitors, causing erosion and detracting from its beauty. Such conditions failed to meet visitors' expectations and overall experience of the park, highlighting the need for improved maintenance and upgrades.

In 1988 the nonprofit Yosemite Association launched its Yosemite Fund to raise needed moneys to support renovation projects throughout the park. The federal government's continued failure to adequately fund the parks, had led the association to act to save the park its members so loved by garnering donations to help the Park Service repair the park's scenic overlooks, manage its wildlife, and build new walkways. The collaboration with the Park Service yielded results, including raising $13.5 million to rebuild the Lower Yosemite Trail in 2005. While successful, the fund underscored the Park Service's increasing reliance on funding from its partner organization, as it struggled to address needs less appealing to donors.[30]

NORBERT AND EDITH Fischer had hoped to bring their daughter Lori to see Yosemite in the summer of 1990. But for the first time in its hundred-year history, the national park was closed. Fire engulfed much of the park's western boundary, forcing the closure of the Arch Rock Entrance and the evacuation of some 10,000 from the Yosemite Valley. "I'm a lot disappointed because I read a chapter about it in my social studies book," the ten-year-old later told a reporter on the Mission Viejo family's change in plans. Rather than Yosemite, they chose to visit Sequoia National Park instead. Norbert promised his daughter that they would try to see Yosemite, just not that summer. "Hopefully next year," he vowed.

Lighting strikes had triggered twenty-seven fires throughout the Sierra Nevada that summer, including two near Yosemite. The larger of the two, named the A Rock Fire after the Arch Rock Entrance, had blown past fire lines and raced toward the communities of El Portal, Crane Flat, Big Meadow, and Foresta. As smoke inundated the Yosemite Valley, the Park Service closed the park for the first time in its history and ordered visitors to leave. Yet the closure of California Highway 140 stranded hundreds in the valley. Many slept in their cars, unable to make it back to their campsites, or, having just come to the park for the day, were now trapped behind the fire line. "There were lines and lines of traffic trying to get out of the valley," Joe Salazar later recalled of the chaotic scene as hundreds of cars sought to escape the fire.

Among those stuck were Rachel Chang and Michael Zheng of Fremont, California. The couple had slept in their car in Yosemite Valley for two nights after the road to their campground was closed. Finally escorted out of the park by rangers, they stopped at the southern entrance gate to ask how they could retrieve their camping gear and clothes they had left behind at Crane Flat campground. A ranger took down their information, telling them "sit tight and keep waiting."

As Chang and Zheng were escaping the Yosemite Valley, firefighters were battling the largest fire in Yosemite's history to that point. Residents of El Portal and Foresta hastily evacuated as the conflagration threatened their homes. Shifting winds saved El Portal but, sadly, pushed the blaze toward Foresta, burning virtually the entire community. Further south, the Steamboat Fire threatened the Yosemite West inholdings and Badger Pass and came within two miles of the grove of giant sequoia. The arrival of Type I and Type II incident management teams, combined with wetter weather, helped extinguish both blazes, but not until twenty-eight structures and some 15,000 acres had burned.[31]

Debate over the Park Service's response to the two fires ignited before firefighters extinguished the final embers. It had been two years since a massive fire within Yellowstone National Park had drawn national attention to federal wildfire policy. Americans had watched the nightly news in horror as a wall of flame burned 793,880 acres in the park, threatening the Old Yellowstone Lodge. "Part of our national heritage is under threat and on fire tonight," CBS news anchor Dan Rather told viewers in early September as fires raged through the park. Wyoming's senior senator, Alan Simpson, blamed the park's superintendent, Robert Barbee, for mismanaging the blazes and called for his resignation. Yosemite's former fire ecologist, Barbee struggled in communicating the Park Service's policy of allowing some fires to burn in order to preserve the park's ecological health—a tough sale to a public who had only seen one of the nation's crown jewels burn.[32]

Since the 1970s the Park Service had embraced the belief that it needed to put fire back onto the landscape in order to bring balance to ecosystems within the parks. To accomplish this, the agency began allowing naturally occurring fires to burn as long as they did not threaten park resources. This "let it burn" policy came under intense criticism in Yellowstone as a series of fires merged and nearly burned down the beloved Old Yellowstone Lodge. The Park Service reversed its policy following Yellowstone, promoting full suppression in order to meet the public's expectations. Even if such expectations ran counter to the science.[33]

In Yosemite, fire ecologists continued the delicate act of conducting prescribed burns, strategically starting small fires to consume the forest floor's accumulated fuels. Their aim was clear, to mitigate the risk of catastrophic wildfires that threatened to engulf the park. Smoke curled into the sky, drawing complaints from visitors, yet those who understood the ecological tapestry recognized these controlled blazes as essential to revitalizing the park's forests. But to the media and the broader public, the A Rock and Steamboat fires seemed to cast doubt on this policy. If prescribed burns were intended to stave off the threat of massive wildfires, one might well ask, why did Foresta burn?

The answer was far from simple. By the summer of 1990, drought had gripped the Sierra Nevada, leaving the woods parched and vulnerable. "Nature did not give us a chance on this one," remarked Acting Forest Supervisor Patrick Kaunert, continuing, "It would have burned with the same intensity regardless of management policy." Moreover, the A Rock Fire had leaped into the crowns of trees, a phenomenon that spread flames from treetop to treetop, bypassing the ground entirely. Such truths often fell on deaf ears amid a public roused by the devastation of their cherished landscape.[34]

Wildfires continued to encroach upon the park. Driven by drought and high winds, the Rim Fire of 2013 burned 255,000 acres, about 77,254 of which lay within Yosemite's borders. The following summer, the El Portal fire burned an additional 4,689 acres. These conflagrations underscored not only the far-reaching effects of climate change but also the sobering reality that no single policy could both restore fire to the ecosystem and prevent all disasters. It is a lesson that the public, grappling with the ever-growing scale and duration of wildfires across the American West, was slowly beginning to learn.[35]

VISITOR NUMBERS CONTINUED to surge throughout the 1990s, cresting the four million mark by 1996. Yet not all Americans enjoyed equal access to the park. Despite the nation's shifting demographics, the number of African Americans and Latinos visiting the park remained a mere fraction of the park's total visitor numbers. Professor of public policy at California State University at Bakersfield Jack Goldsmith noted this fact in the National Parks Conservation Association's monthly magazine, *National Parks*. Writing that of the millions who visited Yosemite in 1993 only 0.4 percent arriving by car were African American, Goldsmith castigated the Park Service for its failure to make minorities feel welcome in the national parks. While the reasons were many for this, he noted, they all boiled down to the uncomfortable truth that all national parks had historically not included minority voices. From the Park Service's early history of catering to affluent white tourists to the nation's moral stains of slavery and segregation, the national park narrative of preservation and enjoyment never included Brown and Black faces.

Moreover, conservation groups' focus on preserving pristine nature following World War II had had the ill effect of narrowing access to the parks. Groups such as the Sierra Club's focus on wilderness ignored the nation's racial strife, reinforcing the image that the environment was not a civil rights issue. And despite the surge of visitors throughout the park system over the past four decades, "Minorities simply do not see themselves mirrored in Park Service employees," Goldsmith wrote. The Park Service's dearth of minority rangers reinforced the belief among many minorities that they are not welcomed in the national parks.[36]

Responses to Goldsmith's article seemingly confirmed this belief. In a letter to the *National Parks* editors, Guy Lucier of Deleon Springs, Florida wrote, "Bringing more minorities into the parks would probably raise the crime rate when the rangers are being forced to spend more of their time in law enforcement than ever before." Fellow reader Joseph Lange of Grand Junction, Colorado agreed, writing, "It is obvious that the activities and environment of the National Park System, which many of us deeply appreciate, are not of great interest to most blacks and Latinos." While it could be argued that such views sat outside of the mainstream, the fact that the magazine published three such letters with none supporting the points made by Goldsmith underscored the tensions over who belonged within the parks. Yet, as the nation's demographics continued to change, so did the answer to this question.[37]

Shifts in visitation demographics, however, lagged that of the nation. A 2009 Park Service study noted that the overwhelming majority of the park's visitors remained white. Asian and Latino visitors had increased to nearly a quarter of the total, while Native Americans and African Americans each comprised less than 1 percent of the visitor base. In reviewing the report, Texas A&M University professor Jim Gramann observed, "The demographic face of America is not reflected in national park visitation, with a few exceptions." Visitors to larger Western parks like Yosemite were predominantly non-Hispanic white, highly educated, and affluent—a reality not lost on Yosemite ranger Shelton Johnson.[38]

One of the Park Service's few Black interpretive rangers, Johnson started his career at Yellowstone National Park in 1987 before moving to Yosemite. He gained national fame starring in the 2009 Ken Burns documentary series *The National Parks: America's Best Idea*, becoming the face of Yosemite to many. Growing up in a military family, Johnson spent the first six years of his life bouncing from California to England, Germany, Kansas City, and finally Detroit. A talented musician and writer, he enrolled in an MFA program in creative writing after graduating from the University of Michigan and a stint in the Peace Corps. After his first year in the program, Johnson began seeking quiet places to write during the summer. His roommate handed him an application for a seasonal job washing dishes at the Old Faithful Inn in Yellowstone National Park. Johnson filled it out and was soon moving to Wyoming. He fell in love with Yellowstone, spending every free moment hiking and camping in the park. Johnson returned the next summer as a summer ranger, and he eventually became a full-time naturalist and interpretive ranger.[39]

After spending several summers in Yellowstone, he moved to Yosemite. There, while working on a ranger talk in the park's research library, Johnson stumbled across a photograph of mounted Ninth Cavalry troops, an all-Black regiment known as the Buffalo Soldiers, patrolling the park. It was a revelatory moment. He felt an immediate connection to the uniformed men in the photograph, and he believed that other African Americans would share that connection to those Buffalo Soldiers and to Yosemite. Inspired, Johnson spent the next several years digging into the Buffalo Soldiers' history, within both Yosemite and Sequoia National Parks, crafting interpretive talks and constructing a cavalry uniform to reenact a composite of Ninth Cavalry veterans for visitors. The lanky ranger with a musician's hands quickly became one of the most beloved rangers within the park. His talks captivated audiences small and large. Moreover, he became a vocal advocate of the park's power in accepting all who wished to experience the outdoors.

Tales of Johnson's work caught the ear of the Ken Burns production team. Johnson's skill as a master storyteller, passion for the history of the Buffalo

Soldiers, and own personal narrative captivated the documentary's audiences, and propelling him into national fame. Soon after the series' release, visitors would come up to him and declare, "You're a movie star. I saw you in the video." Johnson would flash his now famous grin, shake their hands, and humbly reply, "There's a decided discrepancy in income between park ranger and movie star. I can tell you most definitively that I'm not a movie star." Yet his newfound fame provided Johnson a platform to further his call that the national parks were for everyone.[40]

The following summer Johnson wrote celebrity television personality Oprah Winfrey, inviting her to come to Yosemite. "Every year, America is becoming increasingly diverse, but that diversity is not reflected in the national parks, even though African Americans and other groups played a vital role in the founding of national parks," Johnson observed, before asking, "If the national parks are America's playground, then why are we not playing in the most beautiful places in America?" Winfrey agreed and enlisted her friend and fellow television personality Gail King to surprise Shelton by visiting Yosemite.[41]

After buying a tent and camping gear, most of which they never used, Winfrey and King led a production team into Yosemite, surprising a flabbergasted Johnson in the Mariposa Grove parking lot. Shelton took the pair on a tour through the Mariposa Grove, before they took in Glacier Point. That evening Winfrey and King camped in the Lower Pines Campground. Johnson joined the pair to give a ranger talk on the role of Buffalo Soldiers in the national parks. After a restless night of sleep in their rented trailer, Winfrey and King spent the next day learning to fly fish and taking a horseback ride through the valley. All of it was captured on camera, later becoming two episodes of Winfrey's popular television show.[42]

Reflecting on the celebrities' visit, Johnson noted that he had gone from seeing African American faces once a week or so, to seeing them on a daily basis after the show aired. Such anecdotal evidence underscored the gradual increase of people of color within the park, but challenges remain in diversifying the national parks. As Johnson noted a decade after Winfrey's visit, there is still work to be done.[43]

A RARE TROPICAL storm pounded the Sierras with rain throughout New Year's Eve, 1997. The warm rains melted the heavy winter snows on the mountains, causing flooding across the region. In the Yosemite Valley, water cascaded down its waterfalls, causing the usually docile Merced River to spill over its banks in a torrent not seen for more than forty years. Sue Beatty, a restoration ecologist living in the valley at the time of the flood, recalled, "Yosemite Falls was thundering so loud that our windows were shaking in the house, and we were a half mile away from the waterfall. It was just running!" Across the valley, Stan Wacht watched in awe as the water level quickly swallowed Camp Curry, "By the time I boxed my stuff up, the water was up to my knees. I slept on the floor of the cafeteria with about 150 other employees and guests."[44]

At its peak the flood reached 24,600 cubic feet *per second*. Over four days floodwaters washed out large sections of U.S. Highways 120 and 140, destroyed concessioner employee housing, and wreaked havoc on the park's bridges, trails, and campgrounds. Nearly 2,100 people became trapped in the valley's hotels and lodges. By the end the flood was the largest and most destructive in the park's history, causing an estimated $178,000 in damages, and closing the Yosemite Valley for nearly a month. Photographs taken by park employees showed the flood's massive destruction—the Lower River Campground completely underwater, whole sections of Highway 140 washed away, and debris jammed against the Singing Bridge. On surveying the damage to the Lower Pines Campground, park spokesman Geoff Green told the *San Francisco Chronicle*, "It's likely that in this area, the river will keep it." In all, roughly half of the valley's campsites were destroyed, many having been located in the flood plain.[45]

Farther upstream, the flood had swept away the Church Bowl. Built in 1920 this four-hundred-seat amphitheater had been a gathering place for generations of churchgoers. Now it was gone. Nestled in a quiet corner of Yosemite Valley, between the visitor center and the Ahwahnee, the Church Bowl had been created out of necessity, a larger venue to accommodate Sunday services in the early twentieth century. The quaint New England–style Yosemite Valley Chapel could no longer hold the growing summer crowds, so the local nondenominational Protestant congregation began holding services in the natural amphitheater. The Church Bowl became a permanent fixture by 1933, when the Civilian Conservation Corps added several rows of benches. Every summer weekend hundreds of faithful would gather in the amphitheater, celebrating both their faith and the beauty of the park.

By the summer of 1960, an estimated one thousand people attended services each weekend throughout the summer. Among those were Church of Christ congregations from around California, who convened for a week in the Yosemite Valley every July since the 1940s for their Yosemite Family Encampment. Members attended lectures, conducted Bible study, and enjoyed the valley's natural beauty. What had begun as a small gathering quickly grew into one of the most popular weeks of the year. Thousands of church members descended upon the valley every summer over the coming decades, filling its campgrounds and lodges at the peak of the summer season.

From its earliest days, Yosemite has played a spiritual role for many of its visitors. To these faithful, the Yosemite Valley represents the holiest church of American nature. As historian Mark Stoll noted, it was the Rome of reformed nature pilgrimages. Protestantism played a critical role in Americans' understandings of nature and, by extension, Yosemite. Such views remained in place throughout the first half of the twentieth century, though few practiced their faith in the outdoors. That all changed by the middle of the twentieth century as church camps drew thousands into the outdoors, fundamentally altering how American Christians felt and experienced nature. The Church of Christ's Yosemite Family Encampment reflected a much broader trend of celebrating faith in nature. The nondenominational, and socially conservative, association of Protestant churches embraced the idea of retreating into nature to find spiritual renewal and fellowship with like-minded Christians.[46]

The Family Encampment grew into the largest Church of Christ annual gathering on the West Coast. Attendees came not only to pray together but to reconnect with friends. "This is like one big extended family, like a big homecoming for us," attendee Corky Cowart told the *Merced Sun Star* in the summer of 1990. A construction worker from Vacaville, California, Cowart had attended the Family Encampment for nearly a decade by that point. Both he and his wife were among the thousands who crowded into the Church Bowl, many bringing their own folding chairs. "They are known as 'our lady of the lawn chair,'" said the assistant manager of the Ahwahnee. "They're nice people, they're not rowdy or loud," he told a reporter in his backyard near the path to the amphitheater.[47]

Most shared a similar view of the Encampment. Rangers frequently remarked that the group was well-mannered and used chartered buses to travel around the valley, easing traffic congestion. However, the increasing number of attendees was becoming a growing concern for the Park Service. "No matter how well they do, it is unreasonable to have those numbers in a national park," argued Yosemite's chief ranger, Roger Rudolph. He pointed out that the large crowds were diminishing the experience for other visitors. Additionally, attendees had found a way to circumvent the park's ban on reservations for large groups by making individual reservations for the following summer before leaving the park. By 1990 the sheer size of the group, regardless of their good behavior, was starting to present a significant issue.[48]

Two years later the valley's campgrounds saw over 600,000 visitors, prompting the Park Service to begin limiting the number of campers allowed in the valley. As part of the Yosemite Valley Plan, these restrictions reduced the number of campers within the valley. By 1995 the number of worshippers at the Church Bowl declined due in large part to the Park Service's efforts to reduce numbers within the valley. Then came the flood. In just one day, the benches and stage were swept away. As the waters receded, the area emerged as one of the few viable spots for a temporary picnic area in the valley. Even as the park began rebuilding, the Park Service showed no interest in restoring the Church Bowl, leaving it as a picnic ground.[49]

Yet with all the destruction, the flood crystalized three decades of frustrations over development within the Yosemite Valley. Despite the eventual approval of a General Master Plan (GMP) in 1980, which sought to lessen visitor impacts on the valley, visitor numbers continued to overwhelm the park. This was combined with a steady decline in federal funding, creating a significant maintenance backlog. Park administrators faced widespread criticism for their failure to lower visitor numbers and maintain the roads, parking lots, and campgrounds groaning under the growing use. In 1987 the NPS proposed replacing the Sentinel Bridge, noting the historic stone bridge was structurally unsafe for highway loads and that chunks of concrete had already begun to break off, exposing the reinforcement of its steel trusses. But despite its poor condition, repairs to the Sentinel Bridge languished.[50]

Other goals set by the 1980 GMP also fell victim to politics. In 1989 park superintendent John Moorehead dismissed several components of the GMP, stating they were simply too ambitious to achieve. Environmental groups grew increasingly frustrated over the Park Service's lack of progress in reducing the number of visitors to the Yosemite Valley, placing them at odds with the general public, who understood spending a night in the valley to be their birthright. Congress had designated the Merced River, from its headwaters in the Yosemite National Park Wilderness to its impoundment at Lake McClure, as a Wild and Scenic River, requiring the park to prepare a plan to lower visitor impacts on the river.

Signed into law in 1968, the Wild and Scenic Rivers Act (WRSCA) reflected the era's preservationist ethos. Like its sister legislation, the Wilderness Act, the WRSCA required all rivers designated as wild and scenic be preserved in their free-flowing condition "for the benefit and enjoyment of present and future generations." In seeking to balance the law's preservationist obligation with demands for increased visitor access to the Merced, the Park Service finally proposed a master plan for the river's management—a plan that was never released to the public. Furthermore, funding for the plan evaporated, leaving the park in the precarious position of violating federal law but with no means to comply.[51]

By wiping away nearly a century's worth of development within the valley, the flood presented what many in the Park Service saw as a perfect opportunity to finally realize the long-held goals of reducing traffic, restoring ecosystems, and making capital improvements to the park's deteriorating infrastructure. Environmentalists agreed. Congress's appropriation of $178 million to repair roads, bridges, and campgrounds gave park administration the money they believed they needed to finally implement the 1980 GMP. "This is really a once-in-a-lifetime opportunity," the Wilderness Society's western regional director, Jay Watson, told the *New York Times* in the flood's immediate aftermath.[52]

But as the Park Service moved forward, repairing the damage in the wake of the flood's torrent and years of neglect, the ceasefire between environmental groups and the NPS collapsed. In response to a court injunction, the Park Service regrouped and launched a more inclusive Yosemite Valley Plan. The plan sought to reduce development within the valley by removing the Cascades Dam, along with three bridges along the Merced River, removing the roads within the Stoneman and Ahwahnee Meadows, and demolishing several buildings, including the historic Superintendent House—a decision that pitted historical preservationists against conservationist groups.

After three years of public comment and debate, Secretary of the Interior Bruce Babbitt presented the finalized plan in a speech to San Francisco's Commonwealth Club. "I have come to understand that you're a cantankerous, eccentric, passionate, irrational, idealistic, quarrelsome, impossible crowd of people," he told the audience on the impossible task of satisfying all interested parties in managing the Yosemite Valley. Babbitt admitted that the park, along with the public, was suffering from "planning fatigue" after nearly two decades of on-again, off-again efforts to meet the goals set by the 1980 GMP. Yosemite's problem, Babbitt argued, was not too many people, but too many cars. And the plan, he asserted, balanced use with preservation.[53]

David Brower lost his battle with cancer just days before Secretary Babbitt announced the Yosemite Valley Plan. Never shy in condemning the Park Service for what he believed to be the agency's desecration of Yosemite, the ailing Brower had penned a missive attacking the plan. Published posthumously by the *San Francisco Chronicle*, Brower's article opened with a litany of offenses, including the damming of Hetch Hetchy and the construction of the Tioga Road. He railed against the Park Service's Merced River Plan, proclaiming it as vandalism. Yet Brower saved his greatest wrath for the Yosemite Valley Plan. "Placing no limit on the number of current visitors who can visit the park at one time is a violation of the Organic Act and a breach of our contract with future generations," he fumed. Rather than build more hotels and pave new parking lots in the valley, he argued, the Park Service must focus on actual restoration rather than just giving lip service to the concept while continuing to develop visitor facilities.[54]

While environmentalists applauded the removal of infrastructure, they argued the plan promoted development within the valley. Seeking to consolidate needed visitor infrastructure to better protect the valley's environment, the agency proposed significantly expanding the Yosemite Lodge and building a three-story employee dormitory on the site of Camp 4, as well as removing several hundred camping sites and parking throughout the rest of the valley to restore them to their natural condition. Environmentalist organizations and climbers swiftly condemned the plan, citing Camp 4's historical importance to climbing and the Park Service's apparent pro-development stance. Others applauded the Park Service's acknowledgment of the need to allow access for all of those who wished to visit the park.[55]

As with the 1980 general plan, public opinion split into two very different narratives. Many voiced their concern that the plan went too far in favoring preservation over visitor enjoyment, while others complained the plan promoted too much development within the valley. "It seems the National Park Service plans to continually make our national treasures less accessible and less user friendly," claimed one commenter on the plan's attempts to reduce visitor impacts by reducing the number of automobiles in the valley. Others agreed. "This is a public park, not a wilderness area," complained another. "The emphasis should be on effective use by people, not on returning the area to pristine, wild use." But not everyone agreed on the definition of a national park. "If people want TV, computer hook ups, swimming pools, tennis courts, hairdryers and golf courses they should go to a resort," asserted another visitor, arguing Yosemite was not a resort, but a park.[56]

The Friends of the Yosemite Valley, a small preservationist group and among the most outspoken opponents of the plan, sued the Department of the Interior three times attempting to have the Park Service cap the number of visitors allowed into the valley. Explaining their opposition to the Park Service's efforts to reduce traffic by removing campgrounds, reducing the number of parking lots, and rebuilding employee housing destroyed by the flood, the Friends' co-director, Greg Adair, argued, "The thrust of and balance of the plan is the development, and the singular intent is development."[57]

Others such as Congressman George Radanovich, whose district included much of the park, argued the plan was far too restrictive in limiting visitors' access to the park—access many had enjoyed for generations. Seeking to ensure visitor access to the valley, the Republican congressman introduced a bill advocating the reconstruction of some 150 camping sites along the Merced River, the addition of more parking spaces, and scrapping the Park Service's shuttle bus system. While locals, most of whom relied on tourist dollars for their livelihood, supported Radanovich's plan, many visitors were skeptical of its practicality. "It would be [a waste] of time and taxpayer money to overturn it," one park visitor told the *San Francisco Chronicle*. "There's too much development in the valley as it is."[58]

With such wildly differing opinions over the future of the Merced River and the Yosemite Valley, it was no surprise when the entire issue quickly became mired in the courts. At issue was how the Park Service should protect the river's "outstandingly remarkable values." The Friends of the Yosemite Valley argued that hard and fast limits be placed on the number of visitors allowed within the valley, while Park Service researchers argued for a more flexible approach of management, by monitoring use and then deciding carrying capacity through a Visitor Experience and Resource Protection (VERP) framework. A series of court decisions muddled the issue even further by requiring the Park Service to restrict use to prevent the degradation of the park's scenic and natural values.[59]

DESCRIBED BY J. D. Whitney in 1868 as "perfectly inaccessible, being probably the only one of all the prominent points about the Yosemite which never has been, and never will be trodden by human foot," Half Dome remained one of the most, if not the most, iconic mountains within North America. Carved by glacial ice, its sheer face and steep slopes made it one of the most difficult peaks within Yosemite to climb. Mountaineer George Anderson first summited Half Dome in 1875, quickly dispelling Whitney's claim by threading a rope through six-inch iron bolts he pounded into the rock. Anderson's route remained in place for decades, allowing others, including Galen Clark and John Muir, to make the climb. But over decades the bolts became too worn and rusted to be safely used, and so in 1919 the Sierra Club installed a section of two cables set roughly a meter apart and suspended through a series of stanchions anchored into the rock that allowed climbers to easily reach Half Dome's summit. All that was required was the ability to make the hike of sixteen miles round trip, a lack of fear of heights, and enough common sense not to stand on top of a granite dome in a lightning storm, where the only way off was down a set of steel cables.[60]

By the mid-1990s an average of just 575 climbers arrived at the base of the cable system daily. Yet that number nearly doubled over the next decade, peaking at more than 1,200 by the 2010s. With the increasing numbers came increasing environmental costs. Hikers trampled plants, eroded switchbacks, and left piles of litter in their wake. The tread of thousands of rubber soles had polished the granite on the cable route to a gleaming sheen, making footing particularly treacherous during rainstorms. Rangers fielded numerous complaints of human waste hidden underneath every sizable rock on Half Dome's broad summit. Seeking bragging rights, and believing the climb to be much easier than it was, hundreds made the hike unprepared for its strenuousness and technical difficulty.[61]

On making his way to the top of Half Dome in 2007, *San Francisco Chronicle* reporter Peter Fimrite watched as two women panicked on the cables and needed help to get down, and a young man wearing Converse All-Stars sneakers stepped outside the cables to get around clumps of slower climbers. "I can't wear boots," he replied when asked by another hiker about his ill-chosen footwear, "I get all clumsy in boots." The young hiker was not alone. Hundreds attempted the ascent wearing sandals and carrying nothing more than a single water bottle. "There are almost always people at the top who have run out of water, are too exhausted to continue, lack the proper footwear or are simply paralyzed by fear," wrote Fimrite on the circus Half Dome had become.[62]

It was no surprise, then, when the number of accidents on the cables increased with the upsurge in visitors crowding to the summit. All told, five climbers fell to their deaths between 2003 and 2011, three in 2007 alone. Two falls were due to climbers slipping on wet granite, but the third was more troubling. By all accounts, Hiorfumi Nohara was enjoying the climb up the cables. The sunny Saturday morning promised spectacular views from the top of Half Dome, and after waiting for more than forty minutes to get on the cables, the Japanese hiker was anxious to get to the summit. But as the thirty-seven-year-old was talking and laughing with his friends, he suddenly slipped. Following directly behind, fellow climber Brian Mott watched in horror as Nohara fell off the side of the mountain to his death. "Nobody could reach out for him," a bereft Mott told reporters. "If they could have, I know they would have."[63]

Nohara's death underscored the very real consequences of overcrowding on Half Dome, leading to calls to rein in the madness. In response, the Park Service issued an emergency interim permit system in 2009, limiting the number of hikers on the cables during the busiest days of the summer. Two years later the agency extended the system to seven days a week and began setting rules as to how to best manage the popular climb. As in the aftermath of the Merced Flood, the question over Half Dome reflected the reality that visitors themselves were the primary obstruction to the occasional contemplation of natural scenes Olmsted had spoken of a century and a half before. And just like the debate over limiting visitor access to the river, disagreements erupted over whether the Park Service should regulate the number of hikers allowed on Half Dome.

Three general camps emerged in the fight over limiting access to Half Dome. One agreed that some limits were most likely needed. A second reasoned that the only way to ensure the climb complied with the Wilderness Act was to completely remove the cables. And a third, albeit smaller, group argued that all visitors had a right to climb Half Dome so no action should be taken.

After studying the issue, the Park Service released a Draft Environmental Impact Statement for public comment in late 2012, offering five alternatives on how to best solve the problem of overcrowding on the cables. Hundreds responded via email and letter, with comments ranging from "I realize that we cannot administer IQ tests when making a reservation, but this hike is not the right hike for groups of . . . overweight, dehydrated, ill-equipped, out of breath, day hikers" to "I agree that some limits must be placed on the number of visitors, and of the choices still on the table, I would choose the 400 limit, but I think 500 is more reasonable." Several compared the atmosphere at the base of the cables to Disneyland rather than a wilderness hike in a national park.[64]

The Park Service settled on restricting the number of hikers to three hundred through a permit system. Half would be issued to those holding wilderness permits, and half would be allocated through a lottery system. Two rangers would be stationed in the Little Yosemite Valley administration camp for enforcement and compliance. Simply casting the struggle over Half

Dome as an example of the tension between preservation and use masked the more complicated issue of how to manage traditional expectations of access with growing legal and ecological limitations.[65]

The designation of Half Dome as wilderness in 1984 meant it must retain its "primeval character and influence, without permanent improvements." Such requirements seemed to necessitate the removal of the cables. However, the establishment of the route forty-five years before the passage of the Wilderness Act made the cables a historic, if not traditional, way of climbing Half Dome. Visitor opinions as to how to best manage the growing crowds at the base of the final climb depended upon their views of whether access or preservation held primacy, a distinction that remained at the heart of the debate over what constituted appropriate recreation within Yosemite National Park. The root of this debate is how to preserve moments of solitude while allowing access to one of Yosemite's most popular places.[66]

MANAGING THE ESCALATING numbers entering the park remained a pressing problem for the park's law enforcement. To manage such huge numbers, rangers took on an increasingly aggressive stance toward even the smallest infractions. Campers were woken in the middle of the night and forced to allow rangers to search their vehicles in efforts to reduce bear incidents. While this was well intentioned, visitors took a dim view of such interactions. Making matters worse, following the 9/11 terrorist attacks, law enforcement rangers were now armed with assault rifles and body armor. As Joseph Rumburg noted in the aftermath of the Stoneman Riot, the sight of armed rangers did not fit the national park image to most visitors. Stories of rousing unsuspecting concessioner employees from their beds on suspicion of being drunk and issuing enormous fines to visitors for comparatively minor offenses have earned Yosemite's law enforcement rangers a reputation for being overly zealous and left the Park Service open to public condemnation.

One of the most egregious examples of law enforcement's overreach in the park was the arrest and imprisonment of Australian tourists Margaret and Andre Visher by park rangers in 2004. Having just enjoyed a meal at the Ahwahnee Hotel to celebrate Andre's sixtieth birthday, the couple was pulled over on suspicion of drunk driving. Andre, who had been driving, blew a .08, the minimum to be considered drunk in California, while Margaret blew a .06. After the two male rangers frisked the couple, both Andre and Margaret were arrested and brought to the park's jail. Asked why she was being arrested, the ranger allegedly told Margaret she was a danger to herself and others. After paying a $2,500 fine, both were released the following morning.[67]

Such reports had become all too common within Yosemite, leading to very vocal criticisms of the park's zero-tolerance policy toward all infractions, no matter how minor, a charge national park officials denied. Such stories do, however, point to the difficult challenges National Park Service law enforcement faces within Yosemite National Park and, by extension, across the entire national park system, as the number of visitors continues to climb. Balancing visitors' expectations of tranquility and friendly rangers with the very real need to manage those ever-growing numbers of visitors remains a contentious issue within Yosemite. "People bring urban problems with them," explained Yosemite ranger Mike Mayer. "When you have 15,000 to 20,000 people bedded down in this valley, it's a small town of transients." The high-profile murder of four women in the late 1990s underscored the increasingly serious nature of crime enforcement within the park. When combined with ensuring the safety of the crowds traveling through Yosemite Valley every summer day, it is no wonder that law enforcement has become one of the most challenging jobs within the national parks. A job made more difficult as crowds continue to overwhelm Yosemite's campgrounds.[68]

Seeking to enforce the park's long-held policy of limiting all visitors to a two-week stay, with just one in the valley, during the entire year, the Park Service began requiring all staying at Camp 4, or Sunnyside, to register. Climbers immediately protested, pointing out that for generations climbers had enjoyed staying at Camp 4 long term. Time, they argued, is needed to master the skills required to ascend Yosemite's walls. Such protests met with little sympathy from park rangers seeking to reduce overcrowding within the Yosemite Valley. In response, many climbers took to finding hidden spots throughout the valley in which to sleep, creating a cat and mouse game between climbers and law enforcement.[69]

DEAN POTTER AND Graham Hunt leaped off Taft Point into the fading evening light. Each wearing a specially made wingsuit, which by stretching a piece of cloth between a wearer's legs and arms and upper torso allowed the wearer to soar through the air with a measure of control, the two BASE jumpers plummeted toward the Yosemite Valley floor some three thousand feet below. The younger Hunt veered left and then right while Potter held his line directly toward a subtle notch on the Lost Brother ridge, the only safe route from Taft Point to the valley floor.

Standing on the point, Potter's girlfriend Jen Rapp watched through the lens of her camera as the men quickly fell out of sight. She then heard a muffled thud, followed by a terrifying silence. Not knowing what had happened, Rapp frantically scrolled through the images on her camera. Seeing nothing in the display, she raced to the valley floor to search for Potter and Hunt. There, she joined Rebecca Haiyne, Hunt's girlfriend, who had planned on picking up the two men after their illegal jump, in a desperate search for the two men. Perhaps they had been arrested, or injured? Maybe they were hiding from park law enforcement? Slowly it became clear that the unimaginable had happened. After disappearing from Rapp's sight, both Hunt and Potter had hit the Lost Brother ridge below at incredible speed. Neither man survived.

Potter's and Hunt's deaths galvanized the decades-long fight between climbers and the Park Service over the ban of BASE-jumping within Yosemite. The sport's proponents argued that the NPS's prohibition of jumping off the valley's sheer granite walls with a parachute was disproportionally arbitrary, particularly with the agency's allowance of rock climbing and hang gliding. Conversely, the Park Service maintained that BASE-jumping had no place within the national parks. "We're not against BASE jumping as a sport," Yosemite National Park spokesman Scott Gediman told the *New York Times*. "But we have to look at the big picture and its appropriateness in the park." Or, as national park ranger Ray O'Neal explained, "We like to think that people come here to enjoy the scenery, and not the spectacle of people jumping." That distinction, between spectacle and scenery, lies at the center of a much longer running debate over what constitutes appropriate recreational use within Yosemite.[70]

An acronym for building, antenna, span, and Earth, the four categories of platforms most typically used by jumpers, BASE-jumping had a difficult history within the park. The first attempt to parachute off El Capitan was performed in 1966 by two skydivers using conventional round parachutes. While the two survived the jump, both ended up in the hospital with multiple bruises and broken bones. A second attempt was not made until twelve years later when a group led by BASE-jumping pioneer Carl Boenish launched themselves off the top of the granite monolith, surprising both park visitors and rangers. Boenish's film of the jump soon inspired other daredevils to make their way to Yosemite. However, the Park Service quickly banned BASE-jumping while it sought a way to manage the new activity. Defining BASE-jumping as a sport similar to hang gliding, which was allowed in a very limited fashion within the park, the agency briefly allowed jumpers back in the park during the summer of 1980. But liability questions, along with jumpers' refusal to adhere to Park Service regulations, brought the experiment to an abrupt halt, and the NPS banned BASE-jumping throughout the entire park system.[71]

Hunt's and Potter's deaths highlighted a much larger challenge facing the park—that of the emergent role of social media. As a professional athlete, Potter earned his living by capturing his adventures online. Millions across the world logged on to watch his videos defying gravity, which his sponsors loved. Investigators had used footage from his GoPro camera in determining the cause of his death. While captivating, Potter's actions, particularly his ascent of Delicate Arch in Arches National Park, often attracted the wrong sort of attention. The feat garnered massive backlash online and led his sponsor, Patagonia, to fire him. Yet Potter's popularity drew views, causing consternation that others would attempt to mimic his exploits.[72]

"People don't come here for solitude. They are looking for the iconic photo," Glen Canyon spokeswoman Maschelle Zia explained in 2018 of visitors taking ever greater risks to photograph themselves in the national parks. "Social media is the number one driver," she argued. Such actions have, sadly, led to injury and even death. Such was the case with travel bloggers Meenakshi Moorthy and Vishnu Viswanath. The Silicon Valley couple fell to their deaths off of Taft Point while staging a photograph meant for their more than 10,000 Instagram followers, joining numerous others who have fallen while attempting to take a selfie on one of Yosemite's majestic vistas.[73]

While these deaths are undoubtedly tragic, engaging in risky behavior for the sake of a photograph in the park is hardly a new phenomenon. Where romantic ideas like solitude help draw some to Yosemite, in truth most come to the park to enjoy being in a beautiful place, camping, hiking, floating down the Merced, shopping at the gift shop, and contemplating Yosemite's sheer scale. As social media has revolutionized how visitors and the park communicate, it has not changed Yosemite's landscape. Like generations before them, visitors continue to express both their awe of Yosemite's wonder and their frustrations over its crowds, lack of services, and management struggles. Rather than capturing their experiences in a journal, letter, or photo album, they now do so online.

Social media also provides a new way for both visitors and the Park Service to connect with Yosemite. Nowhere was this more apparent than the Park Service's 2016 Find Your Park campaign. In celebrating the agency's centennial, Find Your Park harkened to the Park Service's promotional past by connecting visitors to more than four hundred national park units scattered across the country. The slick corporate-sponsored media campaign encouraged visitors to capture their own experiences by tagging their posts, tweets, and chats with #FindYourPark. Like Mission 66 and Director Stephen Mather's efforts before, Find Your Park succeeded in attracting even more people to the national parks, especially larger parks like Yosemite. Moreover, social media has increased communication between visitors and the Park Service. Visitors have come to expect information on closures and weather conditions within Yosemite. In this way, social media has simply built upon generations of visitors' behavior within Yosemite, adding to both our deeper connection with this place and its gradual deterioration.[74]

This phenomenon has played out in the Mariposa Grove. By the end of the twentieth century, the grove shared many of the same challenges facing the valley below. Summer crowds inundated the lower grove, jostling for a better view of giant sequoia, while trams carrying dozens of visitors at $26 a head wound their way along an asphalt road through the grove. A century of heavy foot and tram traffic, along with the diversion of water from the trees' base, had irrevocably damaged the ancient trees' shallow root system. While the Park Service had reintroduced controlled burns to the grove in 1971, often visitors complained of the heavy smoke caused by such fires.

At times, the cacophony of voices, car and bus motors, and the tram drowned out the grove's serenity. Parking remained elusive. And the crowds loitering outside of the gift shop gave the grove an amusement park–like atmosphere rather than the quiet communal feel the Park Service hoped to instill. Combined with the grove's increasingly dire ecological conditions, the crowding led the Park Service to restore the Mariposa Grove. Armed with a $20 million grant from the nonprofit Yosemite Conservancy, and adding $20 million from its own budget, the agency launched its study of how to best preserve the grove and improve the visitor experience in 2011.

Like every management decision within Yosemite, the restoration of the Mariposa Grove was not without controversy. After years of study, the Park Service presented three options on how to move forward. The preferred plan was to move visitor parking and the gift shop to the grove's periphery, construct new restrooms with flush toilets, remove all of the asphalt paths from among the trees, replacing them with dirt and elevated paths, and end all tram service within the grove. Visitors would park outside of the grove and take a shuttle to the trailhead from which they could hike among the sequoia. Elevated pathways would lessen all impact on the trees' fragile root systems while the abandoned historic Washburn Wagon Road, on which the grant's earliest visitors traveled, would be rehabilitated as a pedestrian path. A new trail would be constructed through the lower grove, including a new overlook to the Grizzly Giant. "Yosemite is a place to turn off your cell phone. It's a place where you aren't in a hurry, where there's not noise and you can contemplate nature," said Yosemite's superintendent Mike Tollefson on announcing the park's plan.[75]

Few opposed the parking lot's removal or construction of new elevated paths, but many attacked the Park Service's decision to end the tram. "Your intentions in trying to preserve the Mariposa Grove are appreciated," wrote one commenter on the park's proposed plan, "but you are eliminating the opportunities of many peoples [*sic*] to experience this phenomenal grove." Another commenter noted he had visited Yosemite with his father and brother since the 1950s, and was no longer physically able to make the hike to the Upper Grove. He protested, "Apparently I will no longer be able to visit it in the summer months as well because you've decided that only those with a lot of time on their hands and who are in solid physical condition are now worthy of the journey through the upper and lower groves."[76]

It is easy to understand such sentiments. After all, for generations we have been promised that Yosemite is ours to enjoy. But that is the crux of the problem. How do we square that promise with Yosemite's very real limitations? There is only one Yosemite Valley after all, and it cannot be all things to all people. There are no easy answers. We have to choose between our continued overuse and saving our environmental heritage for future generations. Yet that choice has for too long been framed as a false decision between either preservation or use. The Organic Act charged the National Park Service with both conserving and providing for the enjoyment of the national park's scenery, history, and wildlife—a mandate many see as paradoxical. What if we viewed the act's language not as a contradiction but simply as a mandate? The seeds of this view may have been sowed in the very grove of sequoias set aside more than a century and a half ago.

UNDER A BRILLIANT blue Sierra sky, Superintendent Michael Reynolds and Yosemite Conservancy President Frank Dean shared in cutting a green ribbon, officially reopening the Mariposa Grove. "These trees sowed the seeds of the national park idea in the 1800s, and because of this incredible project it will remain one of the world's most significant natural and cultural resources," Reynolds told the small crowd gathered for the ceremony. After being closed for three years, the grove now allows visitors to walk among the giant sequoias, contemplating their age, while preserving these ancient trees and their habitat as a part of our natural heritage. Not everyone may be able to visit whenever they wish, but that may be the price of preserving the grove's continued existence and our promise to future generations so that they may enjoy walking among the giants as previous generations have.[77]

Epilogue

Granite, Not Gridlock

Few paid much attention to the headlines telling of a new virus sweeping across Asia at first. It felt far away from their daily concerns and future plans. But within months, COVID-19 was all Americans were talking about as the virus plowed across the country. Public health officials scrambled to contain its spread by abruptly closing public spaces, including Yosemite. The park shut its gates for eighty-nine days, the longest closure in its history. By June 2020, visitors were allowed to return to Yosemite, though in dramatically reduced numbers. Seeking to comply with federal and state health guidelines by keeping visitor numbers low, the Park Service implemented a reservation system limiting visitors to 1,700 per day. "It's going to be a different kind of summer," Yosemite's acting superintendent, Cicely Muldoon, noted. And it was.[1]

Those lucky enough to navigate the challenging online reservation system discovered a Yosemite free of the crowds that had come to define the park during the summer months. Not only was parking easy to find but there were no crowds at popular scenic overlooks. Even the air felt just a little cleaner. "It feels like what the world must have been like before humans," remarked Susan Andaloro while walking through the valley. The Calabasas resident had driven overnight to be one of the first visitors to return to the park. But while she reveled in the Yosemite Valley's beauty, signs of the pandemic's impact were visible throughout the park. The visitor center and museum remained closed. While open, the Village Store required customers to pass through a barricade. Only a few of the park's camping spots were available, and the Ahwahnee and Yosemite Valley Lodges were not expected to reopen anytime soon.[2]

Matters appeared to only get worse when heavy smoke from wildfires in the Sequoia and Kings National Forests forced the Park Service to once again close the park. The combined threat of wildfire and the coronavirus felt apocalyptic to many, underscoring the looming sense of dread from the all-too-sudden unpredictability caused by events far outside their control. Fire and disease felt like they were everywhere. Even the sky over the Bay Area turned dark as heavy smoke blocked out the sun. In the end, more than four million acres burned throughout the state, making 2020 the single largest fire season in California's history.

But as the fires were contained, the virus continued to spread. Seeking respite from the ongoing lockdowns and hungry for some sense of normalcy, people flooded into the outdoors. Beaches, hiking trails, and parks quickly became overrun. By the winter Yosemite had lifted its reservation system, but it soon became apparent that the pandemic was not going to loosen its grip on the country any time soon and restrictions would have to be put back into place before the coming busy summer season.

The new reservation system required visitors to purchase reservations online a month in advance, though 20 percent of the permits were available only two days prior to visiting. Lodging and campgrounds remained limited throughout the park, with all walk-in campgrounds and the High Sierra Camps remaining closed throughout the summer. Frustrations quickly erupted over the reservation system's reinstallation. Permits sold out in minutes online. Visitors with lifetime passes questioned why they needed to secure reservations. Didn't their passes already ensure their entrance to all national parks?

Understanding such frustrations, no doubt fueled in some part by the ongoing pandemic, Superintendent Muldoon noted that visitation was already spiking, citing one visitor who had waited in Yosemite Valley traffic for five hours, and in order to keep everyone safe while protecting the park such measures were necessary. "This should be a trip of a lifetime," she told reporters. "It should be about the granite, not the gridlock."[3]

Despite the rancor against Yosemite's reservation system, it worked. In 2020 a mere 2.26 million people visited the park, a number not seen since the early 1980s. "People were happy again," park ranger Emily Dayhoff told me in a brief phone conversation. Crowds were smaller. People could find parking in the valley. The grasses in the valley's meadow grew above Dayhoff's waist. But it was the quiet that struck her the most. "It was weird, you could hear the birds sing instead of the hum of traffic," she said. Having grown up in the gateway town of Mariposa, the Miwok descendant had spent much of her life in and around Yosemite. Now a demonstrator at the museum following Julia Parker's retirement, Dayhoff spent much of her days talking to visitors, who not only expressed their joy at the lack of crowds but wanted to learn more about the Indigenous peoples of the region, as they had for generations.[4]

The Park Service reinstated the reservation system the following summer. While frustrations continued over the rather opaque rules on securing a reservation and paying an entrance fee, those who spent time in the park reveled in its stillness. Roads were empty of cars and buses, allowing pedestrians to wander wherever they liked. Capped at 20 percent occupancy, the campgrounds were less noisy and more pleasant. Wildlife returned in abundance. All in all, the curbing of visitors made experiencing Yosemite far more pleasant and in line with the promise of what the national parks are supposed to be.

Despite the reservation system's success, the Park Service rescinded the system in 2023. As is often the case, economics played a central role in the decision. Yosemite's gateway communities, ironically the very same ones that had protested the national park's creation more than a century earlier over economic concerns, argued that limiting the number of visitors into the park hurt their annual incomes, and they pressured politicians to force the Park Service to end the reservation system. The decision proved to be a mistake as Stephen Contreras and his family found out. Longtime visitors to the park, the Whittier, California, residents grew exasperated at the hours-long wait to enter Yosemite, and after finally finding parking they grew so frustrated at waiting for a shuttle with enough space that they loaded into the bed of the family truck and drove to the nearest trailhead. Rangers soon pulled Stephen over, fining him $150 for having passengers in the back of an open vehicle. The family's experience was indicative of what most visitors to the park experienced that summer, at least those who were able to get in, and this forced the Park Service to once again reassess how to better manage the park.[5]

The reservation system offered a rare live experiment on the effects of limiting the number of visitors into the park. Not only did smaller numbers help the park's flora and fauna recover but it provided a much better visitor experience. As much as the health of Yosemite's meadows, the clarity of its air, and the well-being of its bears are important environmental concerns, so is the experience of this place. Experience is as much an environmental issue as preserving the Mariposa Grove. Two summers before the pandemic, the Park Service completed restoration of the Mariposa Grove. The largest restoration project in Yosemite's history, the restoration of the Mariposa Grove was just one effort to reconcile visitors' experiences with Yosemite's past. That reconciliation reaches beyond Yosemite's natural environment to conversations about our nation's ongoing struggles over our shared past.

Like much of the country, in the wake of the murder of George Floyd in 2020, the Sierra Club became engaged in a broader discussion about race and reconciling its past complicity in perpetuating white supremacy. In a blog post on the organization's website, the Club's executive director, Michael Brune, announced that the 128-year-old environmental organization was taking a moment to reexamine its past and role in perpetuating white supremacy. It would begin to do so by centering the voices of people the Club had historically ignored, and, more controversially, by acknowledging its founders' racist beliefs, particularly those of John Muir. "Beloved by many of our members, his writing taught generations of people to see the sacredness of nature," wrote Brune on Muir's importance to so many. Yet, throughout his life, Muir had maintained friendships with known eugenicists and white supremacists. Furthermore, Muir himself had often expressed racist views on both African Americans and Native peoples, calling them filthy, superstitious, lazy, and even unnatural.[6]

Predictably, the reaction to Brune's declaration was contentious. Many took to social media to either celebrate the Sierra Club's efforts or demand Brune's dismissal. One disgruntled commenter wrote on the Club's website, "My annual membership came due in September. Because of this article, I decided to tear up my membership card and leave the Sierra Club . . . I believe this is management's pandering to the new 'culture cancel' [*sic*]." Other members pointed out that Muir had changed his views on Native peoples over his lifetime, and rather than being torn down he should be emulated. However, both lines of argument missed Brune's central point. By targeting Muir as a symbol of the environmental movement, rather than Muir as the man, Brune sought to send a powerful message—that the idea that exploring, enjoying, and protecting the outdoors can somehow be separated from human affairs relies on the privilege of being able to ignore the messy business of society, power, and race that is entwined in everything. And that the environmental movement needs to reckon with its racist past.[7]

But Muir represents far more than just the Sierra Club or the modern environmental movement. To millions his very name is synonymous with Yosemite. From his first visit to the Yosemite Valley in 1868 to his death in 1914, Muir's advocacy for Yosemite's preservation and the importance of nature's restorative value defined not only our modern understanding of nature but Yosemite's central place within it. His writings gave expression to the feeling of awe millions still experience when standing on Inspiration Point, or at the base of El Capitan, or walking through the Mariposa Grove. Muir perhaps best captured the promise of the national parks in the opening sentence of his 1901 book *Our National Parks*, writing, "Thousands of tired, nerve-shaken, over-civilized people are beginning to find out that going to the mountains is going home; that wildness is a necessity." Such words reflected those of Fredrick Olmsted in 1865 when he noted the "occasional contemplation of natural scenes of an impressive character . . . it not only gives pleasure for the time being but increases the subsequent capacity for happiness and the means of securing happiness." And even the key sentence of the National Park's Organic Act that mandated that the newly created Park Service "conserve the scenery and the natural and historic objects and the wild life therein and to provide for the enjoyment of the same in such manner and by such means as will leave them unimpaired for the enjoyment of future generations." Such language continues to resonate with visitors, helping draw millions to Yosemite every summer.[8]

But like the Sierra Club, Yosemite National Park needs to reconcile with its legacy when it comes to Native peoples. As numerous historians have noted over the past several decades, the idea that sits at the center of the national parks, an idea expressed by those like Muir, excluded the Native inhabitants of Yosemite. Muir's casting of the Miwok people as debased and filthy mirrored the national park's historic view of the region's Native people. From the invasion of the Yosemite Valley by California militiamen in 1851 to the systematic removal of Miwok and Paiute residents throughout the twentieth century, the Ahwahnechee, Miwok, and Paiute families that called Yosemite home have been seen as little more than hinderances to visitors' enjoyment of Yosemite's beauty and needed to either be removed, as they were in 1969, or kept as invisible as possible, as they have been throughout the park's history. That was unless they could perform in some manner for tourists, as they did during the thirteen years of Yosemite Indian Field Days, or in the Yosemite Museum.[9]

Following the eviction of the residents of the New Indian Village adjacent to Camp 4 in 1969, the National Park Service and the American Indian Council of Mariposa County entered into a prolonged discussion over greater inclusion of Miwok peoples in park management. While the park's superintendent, Les Arnberger, agreed to improving communications between the Park Service and the Native community, he refused to allow Natives any control over park interpretation or ensure employment within the park. It was agreed, however, to replace the Indian Village with a new Cultural Center. While the application for the center was finally approved in 1980, ten years after its submission, the Park Service remained focused on the management of natural resources rather than reconciliation with the region's Indigenous inhabitants.

It took another twenty years before the Yosemite Valley Plan included plans for the construction of a traditional village, including a sweat lodge. Negotiations over the village's placement and construction languished for another eight years, until finally, in 2008, an agreement was struck. Yet, as is too often the case with national parks, the agreement nearly collapsed when then-Superintendent Don Neubacher halted construction on the sweat lodge, citing that it did not meet California building code. Neubacher had just arrived in Yosemite the year prior, after serving as the superintendent at Point Reyes National Seashore, and his tone-deaf decision appeared to be a continuation of the Park Service's long history of dispossessing the Miwok and Paiute. After all, the Miwok had built such lodges for centuries before the establishment of the California Department of General Services.

To meet the codes, the Park Service provided $25,000, which, along with donations, provided for hiring a Native American engineering firm to design and build the lodge and allowed the work to be completed in 2017. Neubacher had been a controversial figure during his time at Point Reyes. Growing complaints over his management style at both parks had forced him to resign from the Park Service the year before. His replacement, Michael Reynolds, proved to be a powerful ally for the Miwok. He signed a thirty-year agreement allowing the Miwok to build and use a new village on the site that the Park Service had razed in 1969. While visitors would be allowed to tour the new village, portions would remain off-limits to all except Miwok people. Such reconciliation appeared to open the door to a better future within Yosemite.

The reestablishment of the Indian Village is beginning a larger reconciliation effort in which the national parks, especially those, like Yosemite, deemed as natural parks, can play a central role. As Native legal scholar and attorney Walter Echo-Hawk Jr. wrote in 2009, the Park Service needs to embrace a broader land ethic that strides "toward a more just culture that has adapted to the land and incorporates valuable indigenous knowledge and values of its Native peoples into the social fabric." This will require a shift in how we think about our national parks. We will have to, in the words of environmental historian Mary Mendoza, think outside of the wilderness box that has dominated our understandings and, more importantly, management of parks like Yosemite.[10]

To reconcile not only with Indigenous peoples but also with the natural world, we will need to reconsider our role in Yosemite. As ecologist Garrett Hardin noted in 1968, there is only one Yosemite Valley. And so we must take our stewardship of this remarkable place more personally. Perhaps, as demonstrated by the reservation system, we need to limit our visitation. Not only to conserve Yosemite's natural wonders but to preserve our shared experiences of those wonders. Experience must become a part of how we measure the environment. To live up to the ideals of national parks may mean that standing on Glacier Point is a truly once-in-a-lifetime event. This is particularly difficult as we also seek to open the experience of visiting Yosemite to those often ignored by history. If we are truly going to share Yosemite with everyone then we need to make space for everyone.[11]

Second, the Park Service needs to undergo its own sea change in how it views its mission and the parks it manages. This means the agency will need to focus as much on the visitor experience as it does on preserving the natural environment. Many disagree with such a view, arguing that the Park Service must manage national parks, particularly those identified as natural parks, as wilderness. This view ignores the fundamental purpose of the parks as laid out in the Organic Act, but, more importantly, it ignores how the millions of visitors who come and enjoy not just Yosemite, but all of the national parks, understand and value the parks.[12]

Finally, to accomplish this goal, the national parks must be fully funded and staffed. This goal has remained elusive since the establishment of Yellowstone National Park in 1872. From that moment, Congress has underfunded the parks, except for two brief moments during the New Deal and Mission 66. But since the 1980s, a fair number in Congress, from across the political spectrum, have embraced privatizing essential services within the national parks. This has left the parks chronically understaffed, and visitors underserved. One of the primary drivers of Yosemite's summer reservation system is its severe lack of personnel. As visitor numbers have risen over the past quarter century, the number of rangers has declined precipitously. According to the National Park Conservation Association, staffing throughout the National Park System has dropped by 20 percent over the past decade, while visitation increased by 16 percent. This decline is largely due to budget cuts and has led to an overworked workforce, and increasing visitor frustrations over access and interpretation.[13]

It will take more than a columnist railing against Congress's negligence to humiliate politicians into adequately funding the Park Service today. The proliferation of social media and the twenty-four-hour news cycle typically drown out such calls. The push to privatize the park has become accepted by both political parties, stifling any discussion on increasing park staffing. One example is the Office of Personnel Management's mindboggling, and ruinous, 2016 ruling that seasonal park employees cannot work more than 1,039 hours anywhere in the National Park System to remain eligible for future federal employment. This Obama-era decision gutted Park Service staffing, which is reliant on seasonal employees more than any other federal agency. But the 1,039 decision paled in comparison to the magnitude of the attack on the parks by the Trump administration.[14]

As I finish this book, in the fall of 2025, the state of the national parks is in chaos. At the start of his second term, President Donald Trump rescinded the offers to thousands of Park Service seasonal hires. Days later, he fired all probationary Park Service employees without warning. In protest, six now-former National Park Service employees unfurled a 30-by-50-foot upside-down U.S. flag from the top of Horsetail Falls. The group released a statement on their protest, claiming that the nation's public lands were under attack, and "Firing 1,000s of staff regardless of position or performance across the nation is the first step in destabilizing the protections in place for these great places." Although the courts rescinded the order, the damage was done. The extraordinary, and by all accounts illegal, firing of Park Service personnel has led to concerns over whether the parks can remain open through the coming summer, let alone fight wildfires, or provide emergency first response, or even unlock bathrooms.[15]

Enraging as such actions are, in many ways they echo the past. I have kept returning to Wallace Stegner's 1983 essay "America's Best Idea" throughout writing this book. Written in protest of Secretary of the Interior James Watt's assault on the national parks, Stegner lamented his fellow citizens' choices at the ballot box, especially when it came to the state of our national parks. Like Stegner, I am both a critical admirer of the great American public and deeply cynical of their election of representatives who push bills that, sadly, confuse corporate interests with the public's interest. In those moments, it is good to remember Stegner's famous words on the true meaning of national parks like Yosemite. That they are "absolutely American, absolutely democratic, they reflect us at our best rather than our worst. Without them millions of American lives, including mine, would have been poorer." These lines remain as true today as they were when Stegner wrote them.[16]

Yosemite will enthrall future generations of visitors in ways both familiar and unimaginable. But we have to work as past generations have to fulfill the Organic Act's promise that Yosemite will remain unimpaired for future generations' enjoyment in experiencing this incomparable place.

Acknowledgments

I want to start by thanking the brown-haired girl who, as a child, spent summers exploring the Yosemite Valley with her brother on their bikes. Years later, after patiently listening to me work through my ideas for this book, she insisted I write about those summers and her Yosemite. You were right as usual, Leisl. Thank you.

Every book has its own history, and this one is no different. This volume began some years ago, when my graduate mentors Andrew Kirk and Hal Rothman hired me as the project lead at the University of Nevada, Las Vegas, for what became the *Yosemite National Park Administrative History*. Over the course of two summers, Yosemite's archivist Paul Rogers opened the park's archives for me to scan every file I could lay my hands on. Yosemite's past came alive in that climate-controlled room at the park's headquarters in El Portal. The thousands of memoranda, letters, and other documents I digitized those two summers became the backbone of the administrative history and, later, this book. After Hal's death in 2007, Lincoln Bramwell took on the herculean task of completing the administrative history, revising my two admittedly rough chapters and writing four others. Through the process, we became fast friends, and now, happily, neighbors.

Char Miller and Matt Becker approached me some years later about taking on writing a book on one of the most beloved parks. I quickly said yes, before fully realizing the enormous challenge of telling Yosemite's history. In 2015 I received the E. Peter Mauk Jr./Doyce B. Nunis Jr. Fellowship, which paid for a month of research at the Huntington Library. There is possibly no other archive as welcoming as the Huntington, whose staff, expansive gardens, and art galleries are unmatched. Particularly, I want to thank Peter Blodgett, who helped me navigate the Huntington's collections.

I worked in two different departments while writing this book. To my friends and colleagues at the University of Northern Iowa, thank you for your generosity and friendship. Know that you are all missed. My former provost, Jim Wohlpart, sent Leisl and me to Yellowstone for two summers as American Association of State Colleges and Universities fellows and was a much-needed mentor during my time at UNI. As luck would have it, I joined another generous department at Colorado State University. My colleagues Jared Orsi, Ann Little, Andrea Duffy, Derek Evert, and Doug Sheflin each read a chapter of the manuscript. A special thanks to Sarah Payne, Jessica Jackson, and Nicole Archambeau for their support and friendship. I am forever indebted to Allison Pine and Avery Cole, both of whom went well beyond the call of duty, and whom I am proud to count as friends. And to Robert Gudmested, who always had time for a quick talk, especially about the 1860s, and a word of support as I drafted the earlier chapters.

Several others have supported and inspired me along the journey that has been this manuscript. After my first year of graduate school at UNLV, Michael Duchemin, Rob Nickle, and I spent a week road-tripping through California. It was my first time seeing Yosemite, including a tour of the collections at the Yosemite Museum set up by Mike. It was an amazing introduction to the park and sent me down the path that led to this book. My friends Christian Harrison and Kent Blansett read versions of the manuscript, with Kent going above and beyond by offering handwritten comments on the entire manuscript. Thank you both, my brothers, for your friendship. I met Jay Taylor at my first academic conference as a graduate student. Since then, he has become both a mentor and dear friend. He has provided a needed sounding board and encouragement throughout my writing of the book. Patrick McCray and David Vail both helped me step away from my computer screen and get outdoors when I needed it most. Thank you both for your friendship and shared adventures. Over the years, I have presented portions of the manuscript to both public and academic audiences. The questions and comments of audiences at the Western History Association and the American Society for Environmental History have all improved my thoughts about Yosemite and its visitors. Michael and Valerie Cohen kindly offered their experiences in Yosemite, and I have been honored to count them as friends.

I want to offer a special thank you to my students at both UNI and CSU, many of whom have helped me develop my ideas by sitting through lectures, crafting keen papers on the national parks, and even sitting down with me to tell me their stories of working in Yosemite. Thank you to Yosemite Rangers Yenyen Chan and Emily Dayhoff for taking the time to discuss your research. A special thanks to Mike Wurtz, head of Special Collections and Archives at Pacific University, for providing clear and insightful answers to my questions about John Muir. I also appreciate Steven Johnson, who generously allowed me to use his illustrations and shared his experiences in Yosemite, and Meg Partridge, who shared stories about her father's work in the park. Dick Orsi graciously read a late version of the manuscript and provided both keen insight into the early creation of the Yosemite Grant and enthusiastic praise. Jamie Lewis published an earlier version of the Stoneman Meadow Riot in the *Forest History Today*, which not only helped me sharpen my ideas but garnered much appreciated public feedback.

I completed writing this book at the National Humanities Center in Durham, North Carolina. My time there writing, thinking, and sharing lunchtime conversations with some of the most remarkable people I've met was made possible by the Robert F. and Margaret S. Goheen Fellowship. To my fellow NHC fellows, whose good company and thoughtful commentary helped put the final touches on the manuscript. In particular, I want to thank Frank Shovlin, Mark Cruse, Emily Hobson, Ashley Carse, Isabella Gomez, Gabriel Ejaik-Rodriguez, Joe Clark, Deborah Mauskopf, Kim Haines-Eitzen, Sonia Harzard, Julia King, Enjung Kim, Susanna Lee, Amy Lonetree, Mostafa Minawi, John Wood Sweet, and Angela Sun for their comradeship during my time in the North Carolina Piedmont.

A special thanks to all those who keep the lights running at the NHC. Thank you Martha Kelly, Joel Elliot, Tara Stoll, Becky Timblin, and James Getkin for making my time at the center so enjoyable. And a note of gratitude to the center's librarians Joe Milillo, Kyle Nguyen, and Shelbi Webb, whose efforts beyond the call of duty in helping me secure both needed books and images made this book. I also enjoyed meeting members of the Triangle Environmental Humanist Seminar while on sabbatical. Thank you, Matt Booker and Tania Munz in particular, for inviting me to the monthly discussions, to Anne Whisnant for sharing her time and thoughts on the national parks, and to the group for workshopping a chapter of this book.

I would like to extend a heartfelt thank you to Bridget Berry, whose wise advice guided me through both completing this manuscript and trusting in Leisl's and my ability to build a series with the University of Nebraska Press. I also want to thank my peer reviewers, Bill Deverell and George Vrtis, for their thoughtful and helpful feedback.

I could not have achieved much of what I have in life without my family. This book is dedicated to my parents. I lost my father, Dale Childers, just before I moved back to Colorado in 2018. From him, I inherited my love of history, reading, and dogs, and I think he would have enjoyed this book. My mom Barbara Childers's unwavering support means the world to me, and I'm grateful to be back home. I also want to thank my in-laws, Ron and Sandy Carr, who introduced Leisl to Yosemite, and Brian and Tricia for their hospitality, company, and humor that summer I worked at the Huntington Library. Thank you to my sister Theresa, a former National Park Service biologist. I treasure our long winding conversations about national parks and their purpose. And to my brother Tim, whose deep love for the outdoors and the West I share.

Finally, thank you all of those who shared their stories of Yosemite with me. I hope I have captured your love for this special place in the coming pages.

Thanks to the editors of the journals where these essays were first published.

"For Public Use, Resort, and Recreation: The Struggle Over Appropriate Recreation in Yosemite," *George Wright Forum* 35, no. 2 (November 2018): 304–311.

"The Stoneman Meadow Riot: Law Enforcement in Yosemite National Park," *Forest History Today* 23, no. 1 (Spring 2017): 28–34.

Notes

Introduction

1. Demars, *The Tourist in Yosemite, 1855–1985*, 1.
2. Jeffrey Jenkins, "Science and the Evolving Management of Environmental Hazards at Yosemite National Parks," *Parks and Stewardship Forum*, September 15, 2022, 540–55.
3. Nearly fifty years ago, geographer Yi-Fu Tuan noted that place is forged through experience, a premise that has served as a compass for generations of geographers and environmental historians in explaining how people's use of, understandings of, and connections to landscapes have shaped history. Tuan, "Place," 152.
4. Spence, *Dispossessing the Wilderness*, 3–5.
5. Muir, *The Yosemite*, 615.
6. An Act to Establish a National Park Service, and for Other Purposes, Pub. L. 64–235, H.R. 15522, 39 Stat. 535, enacted August 25, 1916; Yosemite National Park, "Final Environmental Impact Statement Yosemite Management Plan" (National Park Service, Department of the Interior, 2000), ES-4. In its 2000 Yosemite Valley Plan, the Park Service acknowledged the impacts of overcrowding on visitors' enjoyment of the park, stating, "During peak visitation periods in Yosemite Valley, crowding can diminish visitors' experiences, causing traffic delays, visitor frustration, and impacts to park resources."
7. Runte, *Yosemite*, 1–2.
8. In 1998 historians Roy Rosenzweig and David Thelen identified a significant gap between how professional historians and the public engage with the past. Working with other academic historians, the two conducted a survey asking how Americans of all classes, races, and regions used and understood history. They concluded that while many people held a deep passion for history, their perspectives often diverged sharply from academic interpretations. Rosenzweig and Thelen, *The Presence of the Past*, 5. Other scholars have made similar observations on consumers' role in the production of meaning, including sociologist Colin Campbell and professor of communications Henry Jenkins. See: Campbell, *The Romantic Ethic*; Jenkins, *Convergence Culture*.

9. Thomas, Fowler, and Johnson, *The Silence of the Archive*, ix–xxv; Moss and Thomas, eds., *Archival Silences*, 21.
10. An Act Authorizing a Grant to the State of California of the "Yo-Semite Valley," and of the Land Embracing the "Mariposa Big Tree Grove," Approved June 30, 1864, 13 Stat. 325, in Dilsaver, ed., *America's National Park System*, 11.

1. Making an American Landscape

1. I use the common spelling "Miwok" throughout the book in referring to the Miwok-speaking people who live throughout the Sierra Nevada. That said, it needs to be acknowledged that some nations prefer the spelling Miwuk, or Mewuk.
2. Stoll, "Milton in Yosemite."
3. Bunnell, *Discovery of the Yosemite*, 40–57. The word "Yosemite" came from the Miwok word "yohhe'meti," meaning "they are killers," a name used by Southern Miwok to describe the Ahwahnechee.
4. Solnit, *Savage Dreams*, 219–22.
5. Bunnell, *Discovery of the Yosemite*, 55–65. The California senate committee on forestry reported in 1889 that visitation grew from 167 to 2,590 between 1865 and 1885. Committee on Forestry and the Yosemite Valley and Mariposa Big Tree Grove, "Report of the Commissioners to Manage the Yosemite Valley and the Mariposa Big Tree Grove," 1889, YNPA.
6. Green, *Emerson's Nature and the Artists*, 7–36; Deverell, "Niagara Magnified"; Cronon, "A Place for Stories"; Goetzmann and Goetzmann, *The West of the Imagination*, 143–234.
7. Green, *Emerson's Nature and the Artists*, 7–36; Madley, *An American Genocide*, 186–99.
8. Historian Sven Beckert notes there are few terms in American English as misused, poorly defined, and politically operational as the term "middle class." It is an infuriatingly fluid term that encompasses a wide array of economic, moralistic, and cultural ideas. I use it here to describe the growing number of Americans who embraced aesthetic nature as a marker of their cultural status in the early nineteenth century. Beckert, "Propertied of a Different Kind," 285. Also see: Sears, *Sacred Places*, 4; Bledstein, "Introduction."
9. Greene, *Yosemite*, vol. 2, 20–26.
10. Ostler, *Surviving Genocide*, 381; Madley, *An American Genocide*, 186–99.
11. Greene, *Yosemite*, vol. 2, 32.
12. Huntley, *The Making of Yosemite*, 28–32.

13. Roberts, *American Alchemy*, 1–15; Huntley, *The Making of Yosemite*, 21; Matthews, *The Golden State in the Civil War*, 82.

14. Huntley, *The Making of Yosemite*, 42–43.

15. Ogden, *Yosemite*, 34–36.

16. James Mason Hutchings Scrapbook, 1850–55, YNPRL; James Hutchings, "California for Waterfalls!" *San Francisco Daily California Chronicle*, August 18, 1855, 1.

17. Thomas Almond Ayres, *The High Falls* (June 1855) H 49.8, W 34.7, YNPRL. YOSE 2098, Yosemite National Park; James Mason Hutchings, "The Yo-Ham-I-Te Valley and Its Waterfalls," *Hutching's California Magazine*, vol. 1 (July 1856): 2–7.

18. "Yo Semity Valley," *Mariposa Gazette*, July 12, 1855; James Mason Hutchings Scrapbook, 1850–55, YNPRL.

19. Kurutz, "Yosemite on Glass," 57.

20. Green, *Carleton Walkins*, 86.

21. Carleton E. Watkins, "Scenes in Yosemite Photograph Album, 1870," photCL 409, Huntington Library; Green, *Carleton Watkins*, 87–88.

22. Sargent, *Galen Clark*, 15–38.

23. Sargent, *Galen Clark*, 43–72; James Mason Hutchings to William J. Howard, January 20, 1879, William J. Howard Papers, 1859–1879, Huntington Library.

24. Green, *Carleton Walkins*, 31–43.

25. Green, *Carleton Watkins*, 47–48.

26. Starr, *Americans and the California Dream*, 100.

27. King, *A Vacation among the Sierras*, 35.

28. King, *A Vacation among the Sierras*, 31; Greeley, *An Overland Journey*, 313.

29. "Report of the Commissioners to Manage the Yosemite Valley and the Mariposa Big Trees, 1889–90" (Sacramento: Superintendent of State Documents, 1890), 6–10; Rothman, *Blazing Heritage*, 17–18.

30. Sears, *Sacred Places*, 127.

31. Green, *Carleton Watkins*, 84–87.

32. "Fine Arts," *New York Daily Times*, December 12, 1862, 2; Matthews, *The Golden State in the Civil War*, 148–49.

33. Israel Ward Raymond to Hon. John Conness, February 20, 1864, reprinted in Hans Huth, "Yosemite, The Story of an Idea: An Account of the Changes in the American Attitude Toward Nature That Led to the Creation of Yosemite as a Park," *Sierra Club Bulletin* 33, no. 3 (March 1948): 66; Runte, *Yosemite*, 20–21.

34. Rosenzweig and Blackmar, *The Park and the People*, 42–58.

35. Rosenzweig and Blackmar, *The Park and the People*, 238–59.

36. Roper, *FLO*, 66–232.

37. Diamant and Carr, *Olmsted and Yosemite*, 53; Orsi, *Sunset Limited*, 358.
38. Harriet Errington to Charlotte [Lottie] Errington Field, July 3, 1864, Harriet Errington's Letters and Journal from California, 1864–1865, Papers of John Olmsted Collection, Francis Loeb Library.
39. Roper, *FLO*, 265.
40. Harriet Errington Diary, July 4–August 16, 1864, Harriet Errington's Letters and Journal from California, 1864–1865, Papers of John Olmsted Collection, Francis Loeb Library.
41. Harriet Errington to Charlotte [Lottie] Errington Field, August 10, 1864, Harriet Errington's Letters and Journal from California, 1864–1865, Papers of John Olmsted Collection, Francis Loeb Library.
42. Frederick Law Olmsted to John Olmsted, September 14, 1864, Supra note 83 in Roper, *FLO*, 509; Errington Diary, August 16, 1864; Roper, *FLO*, 269–70.
43. Runte, *Yosemite*, 19.
44. Olmsted, "Yosemite and the Mariposa Grove"; Drabelle, *The Power of Scenery*, 82.
45. Olmsted, "Yosemite and the Mariposa Grove," 143–44.
46. Synopsis of California Statute, 1866, in Letters Received by the Office of the Secretary of the Interior Relating to National Parks, 1872–1907 (Yosemite), RG 79, National Archives, Box 89, 2–3.
47. State of California, *Report of the Commissioners to Manage the Yosemite Valley and the Mariposa Big Tree Grove*, 3–8.
48. Whitney, *The Yosemite Book*, 22–23.
49. Sears, *Sacred Places*, 182–208; Blodgett, "Visiting the 'Realm of Wonder,'" 34–38; Rothman, *Devil's Bargains*, 29.
50. State of California, *Biennial Report of the Commissioners to Manage the Yosemite Valley and the Mariposa Big Tree Grove for the Years 1874 and 1875*, 10–17; Sargent, *Yosemite's Innkeepers*, 12–13; Huntley, *The Making of Yosemite*, 1–3.
51. Huntley, *The Making of Yosemite*, 9.
52. John Muir to Sarah Muir Galloway, September 3, 1873, John Muir Papers, University of the Pacific Library, Holt-Atherton Special Collections.
53. John Muir, "Yosemite Glaciers," *New York Tribune*, December 5, 1871; Samuel T. Farquhar, "Meeting Emerson in Yosemite," *Sierra Club Bulletin* vol. 19 (April–May 1934): 48–55.
54. John Muir to Jeanne Carr, July 26, 1868, John Muir Papers, University of the Pacific Library, Holt-Atherton Special Collections; Email from Mike Wurtz, curator of the John Muir Papers, to author, March 1, 2024.
55. Worster, *A Passion for Nature*, 211; Muir, *Our National Parks*, 512–13.

56. "Report of the Commissioners to Manage the Yosemite Valley and the Mariposa Big Tree Grove," 1889, YNPA.
57. William Hammond Hall, "To Preserve from Defacement and Promote the Use of the Yosemite Valley," report of state engineer, appendix to *Report of the Commissioners to Manage the Yosemite Valley and the Mariposa Big Tree Grove, 1885–56*, 17.
58. Wrobel, *Global West, American Frontier*, 48–82.
59. Sargent, *Wawona Yesterdays*, 26–27.
60. "The Yosemite," *San Francisco Daily Examiner*, July 29, 1888; Isaiah Choynski, "Letter to Secretary of the Interior," November 12, 1890, in Letters Received by the Office of the Secretary of the Interior Relating to National Parks, 1872–1907 (Yosemite), RG 79, National Archives, Box 89; "Letter from the Secretary of the Interior in Response to Senate Resolution of September 22, 1890, Relative to the Supposed Spoliation of Lands Granted to the State of California in 1864," Senate Ex. Doc. No. 22, 52nd Cong., 2nd Sess., in Letters Received by the Office of the Secretary of the Interior Relating to National Parks, 1872–1907 (Yosemite), RG 79, National Archives, Box 89; Sargent, *Yosemite's Innkeepers*, 4.
61. *Report of the Commissioners to Manage the Yosemite Valley and the Mariposa Big Trees, 1889–90*, 6–10; Rothman, *Blazing Heritage*, 17–18.
62. John Muir, "The Treasures of Yosemite," *Century Magazine* 40, no. 2 (August 1890); John Muir, "Features of the Proposed Yosemite National Park," *Century Magazine* 40, no. 5 (September 1890); Cohen, *The History of the Sierra Club*, 7–8.

2. Making a National Park

1. Newsham to Noble, November 24, 1890, File 2679, Box 89, in Letters Received by the Office of the Secretary of the Interior Relating to National Parks, 1872–1907, National Archives (Yosemite), RG 79, YNPA.
2. "Famine in Oklahoma, No Food, No Work, No Seed," *San Bernardino Daily Courier*, August 26, 1890, 1.
3. Newsham to Noble, November 25, 1890, File 2679, Box 89, in Letters Received by the Office of the Secretary of the Interior Relating to National Parks, 1872–1907, National Archives (Yosemite), RG 79, YNPA.
4. Newsham to Noble, November 25, 1890.
5. Hampton, *How the U.S. Cavalry Saved the National Parks*, 81–94; Jacoby, *Crimes against Nature*, 97–102.
6. Robbins, *Our Landed Heritage*, 310–426; Carstensen, *The Public Lands*, xiii–xxvi; Hays, *Conservation and the Gospel of Efficiency*, 66–90; Kelly, *The Hunter Elite*, 80–125.

7. "An Act to Set Apart Certain Tracts of Land in the State of California as Forest Reservations," HR 12197, U.S. Statutes at Large, Vol. 26, Chap. 1263, 650–52. Quoted in Ise, *Our National Park Policy*, 56.
8. "An Act to Set Apart a Certain Tract of Land Lying Near the Headwaters of the Yellowstone River as a Public Park," March 1, 1872, Enrolled Acts and Resolutions of Congress, 1789–1996, General Records of the United States Government, RG 11, YNPA.
9. Meyerson, *Nature's Army*, 67–83; Jacoby, *Crimes against Nature*, 81–148.
10. Meyerson, *Nature's Army*, 90–91.
11. Meyerson, *Nature's Army*, 96; Wood to Noble, July 16, 1891, Box 89, in Letters Received by the Office of the Secretary of the Interior Relating to National Parks, 1872–1907 (Yosemite), RG 79, YNPA.
12. Greene, *Yosemite Historic Resource Study*, vol. 2, 65–66; "Map of Yosemite National Park" (San Francisco: Dewey and Company, 1891), Box 89, in Letters Received by the Office of the Secretary of the Interior Relating to National Parks, 1872–1907 (Yosemite), RG 79, YNPA.
13. *Mariposa Gazette*, December 20, 1890.
14. "Report of the Acting Superintendent of the Yosemite National Park," August 31, 1891 (Washington DC: Government Printing Office, 1891), 4, in Letters Received by the Office of the Secretary of the Interior Relating to National Parks, 1872–1907 (Yosemite), RG 79, YNPA.
15. "Report of the Acting Superintendent of the Yosemite National Park," August 31, 1891 (Washington DC: Government Printing Office, 1891), 4, in Letters Received by the Office of the Secretary of the Interior Relating to National Parks, 1872–1907 (Yosemite), RG 79, YNPA.
16. Kurutz, "Yosemite on Glass," 71–74.
17. Jackson, *Ah-Wah-Ne Days*, 38; Muir, *The Mountains of California*, 93.
18. Meyerson, *Nature's Army*, 102–3; Spence, *Dispossessing the Wilderness*, 109–11; Report of the Acting Superintendent of the Yosemite National Park to the Secretary of the Interior (Washington DC: Government Printing Office, 1891), YNPRL (970:33 I–1), Historic Resource Study, 13.
19. Joachim Henry Senger and William D. Armes, "Letter from J. H. Senger, Wm D. Armes to John Muir," May 25, 1892, John Muir Papers, University of the Pacific, Holt-Athern Special Collections Digital Archives, https://scholarlycommons.pacific.edu/muir-correspondence/178; Cohen, *The History of the Sierra Club*, 9–10.
20. Articles of Association, By-Laws, and List of Members (San Francisco: Sierra Club, 1892), quoted in Cohen, *The History of the Sierra Club*, 9.

21. Cohen, *The History of the Sierra Club*, 8–11; "Reports of the Secretary of the Interior Relative to Yosemite Park, 1892" (Washington DC: Government Printing Office, 1893), 4, in Letters Received by the Office of the Secretary of the Interior Relating to National Parks, 1872–1907 (Yosemite), RG 79, YNPA.
22. U.S. Congress, House Committee and Subcommittee on the Public Lands, *Western Boundary of Yosemite National Park*, 62nd Cong., 2nd Sess., March 20, 1912, 7, in Letters Received by the Office of the Secretary of the Interior Relating to National Parks, 1872–1907 (Yosemite), RG 79, YNPA.
23. "Abstract of Report by Captain J. S. Stidger, Concerning Yosemite Park Affairs," in Letters Received by the Office of the Secretary of the Interior Relating to National Parks, 1872–1907 (Yosemite), RG 79, YNPA.
24. Robinson to Johnson, October 28, 1892, Robert Underwood Johnson Papers, 1889–1924, Bancroft Library, University of California, Berkeley, Box 9, Folder 672; Robinson to Johnson, February 11, 1893, Bancroft Library, University of California, Berkeley, Box 9, Folder 674, quoted in Jones, *John Muir and the Sierra Club*, 57.
25. Cohen, *The History of the Sierra Club*, 14.
26. Young to Interior Secretary, "In the Matter of the Complaint of John L. Howard, against Colonel S. B. M. Young, U.S.A. Superintendent of Yosemite National Park," November 30, 1896; Goodrich reports to Young, September 18, 1896; John L. Howard et al., separate copies to Daniel Lamont, Secretary of War, and David Rowland Francis, Secretary of the Interior, September 15, 1896.
27. Childers, "For Public Use, Resort, and Recreation."
28. Caroline Van der Burg, *A Trip to Yosemite*, with an introduction by Carolyn Lansden Whittle, Yosemite Association, 19, Huntington Library.
29. Johnston, *Yosemite's Yesterdays*, 8.
30. Hyde, "From Stagecoach to Packard Twin Six," 160; E. P. Washburn to Major Harry C. Benson, June 2, 1907, Cavalry and the National Park Service, Correspondence on Automobiles Entering the Park, 1907–1914, YNPA; W. A. Clark, quoted in Johnston, *Yosemite's Yesterdays*, 14.
31. Sierra Club Board of Directors to Hon. Richard H. Ballinger, Secretary of the Interior, May 5, 1909, Calvary and the National Park Service, Correspondence on Automobiles Entering the Park, 1907–1914, YNPA; Acting Secretary of Interior Frank Pierce to A. C. Shaw, March 14, 1911, Correspondence on Automobiles Entering the Park, 1907–1914, YNPA.
32. Louter, *Windshield Wilderness*. Lord James Bryce, "National Parks—The Need of the Future," *Sierra Club Bulletin* (January 1913): 29; John Muir to Howard

Palmer, December 12, 1912, John Muir Papers, University of the Pacific Library, Holt-Atherton Special Collections.

33. Robert M. Price, Secretary, to Members of the Sierra Club, *Sierra Club Circular*, no. 17 (December 20, 1898): 1; Cohen, *The History of the Sierra Club*, 19.
34. William Colby, Interview by Corinne Lathrop Gilb (October–November 1953), Bancroft Library, Regional Oral History Office.
35. On the new businesses open in the Yosemite Valley between 1900 and 1904, see: Greene, *Yosemite Historic Resource Study*, vol. 2, 350–354.
36. John Muir to Robert Underwood Johnson, March 4, 1890, reprinted in Badè, *The Life and Letters of John Muir*, 237; Cohen, *The History of the Sierra Club*, 18–19; William Colby, Interview by Corinne Lathrop Gilb (October–November 1953), Bancroft Library, Regional Oral History Office.
37. William Colby, "Secretary's Report from May 2, 1901, to May 10, 1902," *Sierra Club Bulletin* (May 1902): 147.
38. Muir, *Our National Parks*, 19, 305; Cohen, *The History of the Sierra Club*, 18–19; Worster, *A Passion for Nature*, 371–72.
39. Unknown photographer, "President Roosevelt," in Travel Album of Yosemite, Sierra Nevada Mountains, Yellowstone, and Other Sites, 1903–1905, photo CL 551, Huntington Library.
40. John Muir to Charles Sprague Sargent, March 12, 1903, John Muir Correspondence, University of the Pacific, Holt-Atherton Special Collections Digital Archive.
41. Worster, *A Passion for Nature*, 368.
42. Charlie Leidig's Report of President Roosevelt's Visit in May 1903, in Marvin Koller, "Presidential Visits to Yosemite, Part II," *Yosemite Nature Notes* 38, no. 3 (March 1959): 30–36.
43. William Frederic Badè, "The Tuolumne Caóon," *Sierra Club Bulletin* 5, no. 4 (June 1905): 296. On the importance of Chinese immigrant cooks in the California Sierra Nevada, see: Chan, "Interpreting the Contributions of Chinese Immigrants in Yosemite National Park's History."
44. "Report of the Outing Committee," *Sierra Club Bulletin* 5, no. 3 (January 1905): 240.
45. Chan, "Interpreting the Contributions of Chinese Immigrants in Yosemite National Park's History," 299.
46. Chung, *Chinese in the Woods*, 66–70.
47. Righter, *The Battle over Hetch Hetchy*, 6; Brechin, *Imperial San Francisco*, 13–70.
48. John Muir, "The Hetch Hetchy Valley," *Sierra Club Bulletin* 6, no. 4 (January 1908): 216–17.

49. Woodrow Wilson, December 19, 1913, in Old Central Files Collection, Hetch Hetchy file 1913, Old Central Files, Box 32 of 57, Hetch Hetchy, 1913–1923, YNPA; Ise, *United States Oil Policy*, 334–37; Shankland, *Steve Mather of the National Parks*, 202–4, 211–13. On recrafting of historical narratives, see: Trouilot, *Silencing the Past*, 2; Righter, *The Battle over Hetch Hetchy*, 206–241.
50. Righter, *The Battle over Hetch Hetchy*, 63.
51. Fredrick Law Olmsted Jr. to Mark Sullivan, December 19, 1910, quoted in Sellars, *Preserving Nature in the National Parks*, 30.
52. Horace Albright as told to Robert Cain, quoted in Cain, *The Birth of the National Park Service*, 15.
53. Albright and Albright Schenck, *Creating the National Park Service*, 30–32.
54. An Act to Establish a National Park Service, and for Other Purposes, Pub. L. 64–235, H.R. 15522, 39 Stat. 535, enacted August 25, 1916.
55. Sellars, *Preserving Nature in the National Parks*, 29.

3. California's Playground

1. "Lavelle Girl in California," *Baraboo Weekly*, November 2, 1922, 1.
2. Greene, *Yosemite Historic Resource Study*, vol. 2, 580.
3. McClelland, *Building the National Parks*, 150–51; Shankland, *Steve Mather of the National Parks*, 266–67.
4. McClelland, *Building the National Parks*, 164; Carr, *Wilderness by Design*, 112–13.
5. Meringolo, *Museums, Monuments, and National Parks*, 73–74.
6. Harold E. Perry, "The Yosemite Indian Story: A Drama of Chief Tenaya's People," manuscript, 1949, Huntington Library; Bates, "Native Americans Perpetuating a Wealth of Cultures," 12; Scott, *Yosemite*, 163; Delora, *Playing Indian*, 7.
7. "Yosemite Field Days Full of Incident," *Merced County Sun*, August 11, 1922, 1.
8. Moses, *Wild West Shows and the Images of American Indians*, 168–69; Phillips, *Staging Indigeneity*, 7.
9. Cothran, "Working Indian Field Days," 195; Spence, *Dispossessing the Wilderness*, 116–20.
10. Bates and Lee, *Tradition and Innovation*, 109–10.
11. "Census of Yosemite Indians October 1, 1927," YNPRL (970:33 I–1), "Superintendent's Report for the Month of January 1930," YNPRL; "Superintendent's Report for the Month of February 1930," YNPRL; "Superintendent's Report for the Month of December 1931," YNPRL, 3, 5.
12. "Superintendent's Report for the Month of April 1933," 11, 16, YNPRL; Spence, *Dispossessing the Wilderness*, 121–22.
13. Kiester Jr., *Donald B. Tresidder*, 19–21.

14. National Park Service, *A Sense of Place*, 141; Carr, *Wilderness by Design*, 142.

15. Kiester Jr., *Donald B. Tresidder*, 26.

16. Gabriel Sovulewski, "The Story of Campgrounds in Yosemite Valley," *Yosemite Nature Notes* 16, no. 11 (November 1937): 81–84; Tresidder, "The National Parks," 119.

17. Colton and Dilsaver, "The Hidden Landscape of Yosemite National Park," 38–46; C. G. Thomson, "Superintendent's Monthly Report, June 1930," 2, YNPRL; E. M. Hilton, "Yosemite Valley Sewage Treatment Plant," n.d., typewritten ms. in Box 11, YNPRL.

18. Greene, *Yosemite; The Park and Its Resources*, vol. 2, 581–582.

19. Runte, *Yosemite*, 135–40; *Yosemite Nature Notes* (May 1925). "Superintendent's Report for the Month of July 1927," YNPRL; C. H. Neal, "The Slaughter of the Trusting," *Yosemite Nature Notes* (December 1928); "Superintendent's Report for the Month of October 1929," YNPRL; C. A. Hartwell, "Our Yosemite Bears," *Yosemite Nature Notes* (March 1931).

20. "Superintendent's Report for the Month of June 1924," YNPRL; A. E. Borell, "Yosemite Elk Herd Moved to Owens Valley," *Yosemite Nature Notes* (December 1933).

21. Michel O'Shaughnessy, "O'Shaughnessy Dam Dedication," *Municipal Report* 16 (July 19, 1923) in Righter, *The Battle over Hetch Hetchy*, 152, note 40.

22. Arno Crammer to Acting Secretary of Interior, July 7, 1923, Old Central Files Collection Hetch Hetchy, Box 28, 1923–1927, YNPA; W. B. Lewis to Horace Albright, November 20, 1925, Linda Greene Notes, Box 2, Central Files, Section 1A, Row 7, YNPA; Righter, *The Battle over Hetch Hetchy*, 202.

23. Greene, *Yosemite*, vol. 2, 550–551; W. B. Lewis, Hetch Hetchy Road Opening Commences New Era of Yosemite Development, September 19, 1925, Linda Greene Notes, Box 2, Central Files, Section 1A, Row 7, YNPA.

24. W. B. Lewis to H. B. Hommon, May 26, 1927, Old Central Files, Box 32 of 57, Hetch Hetchy, 1923–1927, YNPA; W. B. Lewis to Director Horace Albright, June 13, 1927, Old Central Files, Box 32 of 57, Hetch Hetchy, 1923–1927, YNPA.

25. Gilbert Jones, Chairman, Legislative Committee of the Pacific Rod and Reel Club of California to W. B. Lewis, June 7, 1927, Old Central Files, Box 32 of 57, Hetch Hetchy, 1923–1927, YNPA.

26. Jones to Lewis, June 7, 1927.

27. Stephen Mather to the Superintendent of Yosemite National Park, June 23, 1973, Old Central Files, Box 32 of 57, Hetch Hetchy, 1923–1927, YNPA.

28. "City Resists U.S. Stand on Hetch Hetchy Roads," newspaper clipping, Old Central Files, Hetch Hetchy, 1923–1927, Box 84 B, YNPA; Stephen T. Mather to Horace M. Albright, August 11, 1927, Old Central Files, Hetch Hetchy, 1923–1927, Box 84 B, YNPA.

29. Congressman Louis Crampton on the Hetch Hetchy Valley, October 5, 1927, Old Central Files, Hetch Hetchy, 1923–1927, Box 84 B, YNPA.
30. John Wosky to Homer Beattie, June 16, 1938, Old Central Files, Hetch Hetchy, 1928–1936 and 1930–1940, acc. 5121, Box 2 of 2, YNPA.
31. Robert O. Binneweis to Mr. Raymond Walsh, Chief Division of Water Rights State Water Resource Control Board, July 20, 1982, Resources Management, acc. 5628, Division Chief Riegelhuth, Hetch Hetchy 1969–1982, YNPA.
32. Superintendent's Report for January 1928 (February 7, 1928), 10–11, YNPRL.
33. Rose, *Magic Yosemite Winters*, 20, and 108; Wrenn, "Playing in the Park," 90–98.
34. National Park Service, "Special Report on Winter Sports Season, 1931–1932," 19. Badger Pass Development Records, Central Files, acc. 5121, YNPA.
35. Donald Tresidder to Colonel C. G. Thomson, August 22, 1935, Badger Pass Development Records, Central Files, acc. 5121, YNPA; Superintendent Thomson to Don Tresidder, August 19, 1935, Badger Pass Development Records, Central Files, acc. 5121, YNPA.
36. E. M. Hilton, Memorandum to the Superintendent, Re: Comments on Letter of May 22 from Dr. Tresidder Regarding Winter Activities, June 28, 1939, Badger Pass Development Records, Central Files, acc. 5121, YNPA.
37. Meyer, "Alpine Experiments," 5–7.
38. Maher, *Nature's New Deal*, 10–11; Orr, "The Civilian Conservation Corps," 3–6; Carr, *Wilderness by Design*, 262–267.
39. Nelson, "Remembering the CCC," 184–191; Email from Dr. William E. Skuban, PhD, Professor and Chair, Department of History, California State University, Fresno, November 12, 2021; Orr, "The Civilian Conservation Corps," 3–6.
40. U.S. Department of the Interior, National Park Service, Yosemite National Park, Fourth Regional Office. Lawrence Merriam, *Report on Flood-Storm Damage December 9–12, 1937*, December 19, 1937, 2, Central Files, Central Files, Flood Storm Damage, 1937–1939, YNPA.
41. U.S. Department of the Interior, National Park Service, Yosemite National Park. Lon Garrison, *Final Report on Yosemite Valley Campgrounds Season of 1939*, November 3, 1939, Central Files, Flood Storm Damage, 1937–1939, YNPA.
42. Frank Kittredge, "Yosemite during the War Years," *Yosemite Nature Notes* 25, no. 5 (May 1946): 76; Ralph Anderson, "Frank A. Kittredge Appointed to New Position," *Yosemite Nature Notes* 26, no. 12 (December 1947): 127–28.
43. Meeting of the Yosemite Advisory Board Held at Yosemite National Park, April 4 and 5, 1942, Linda Greene Notes, Central Files, Section 1A, Row 7; Photograph by R. H. Anderson, YNPRL. Frank A. Kittredge, "Yosemite during the War Years," *Yosemite Nature Notes* 25, no. 5 (May 1946): 74–75.

44. Superintendent of Yosemite to National Park Service Director, November 14, 1942, YNPA.
45. Miller, *Wine, Women, Warren, & Skis*, 14; Anne Goodwin Sides, "The Godfather of Holy Sh*t," *Outside Magazine*, November 2004, https://www.outsideonline.com/1908796/godfather-holy-sht.
46. Memorandum to the Superintendent, March 3, 1943, Central Files, Box 643, Trails 1928–1952, YNPA.
47. Memorandum to the Director, March 4, 1943, Central Files, Box 643, Trails 1928–1952, YNPA.
48. Newton Drury, "The National Parks in Wartime," *American Forests* (August 1943): 37–42; McDonnell, "World War II," 22–25.
49. *History of the United States Naval Special Hospital.*
50. *Yosemite Sentinel* (January 5, 1945), 2; Monthly Superintendent Report, December 1944, 4, YNPRL.
51. Frank Kittredge to E. G. Cahill, December 9, 1941, Old Central Files, Hetch Hetchy, 1937–1942, acc. 5121, Box 34 of 57, Sec. 1A, Row 2, Shelf 56, YNPA; John Wosky, Memorandum for the Superintendent, February 9, 1942, Old Central Files, Hetch Hetchy, 1937–1942, acc. 5121, Box 34 of 57, Sec. 1A, Row 2, Shelf 56, YNPA.
52. Carl Russell to Newton Drury, June 26, 1950, Old Central Files, Hetch Hetchy, 1937–1942, acc. 5121, Box 34 of 57, Sec. 1A, Row 2, Shelf 56, YNPA; Simpson, *Dam!*, 271–75.
53. Burgess, *Trains to Yosemite*, 77; "Machine Tool to Buy Bonds," *New York Times*, February 21, 1945, 26; Alfred Runte, "Yosemite Railroad: Highway of History, Pathway of Promise," *National Parks and Conservation Magazine* 48 (December 1974): 7; Johnston, *Shortline to Paradise*, 28–31.
54. Runte, "Yosemite Railroad," 9.
55. Swan and Roaman, *YV88*; Interview with Christopher Swan, January 26, 2021.
56. Emil F. Ernst, "Preliminary Report on the Study of the Meadows of Yosemite Valley," United States Department of the Interior, National Park Service, Yosemite National Park, California, May 15, 1943, 51.
57. Yosemite Advisory Board Meetings, September 16 to 22, inclusive, 1944, and August 11 to 17, inclusive, 1945, Yosemite Advisory Board Report, 1944–1945, Box #9, Advisory Board Reports (1939–1954), YNPRL.
58. Frank Kittredge, Memorandum, Development in Yosemite Valley, June 25, 1947, Yosemite Development, YNPRL.

4. Let's Open the Parks

1. Bernard DeVoto, "The Easy Chair," *Harper's Magazine* 193 (July 1946): 28.

2. Stegner, *Uneasy Chair*, 290.
3. Bernard DeVoto, "The Easy Chair," *Harper's Magazine* (October 1946): 313.
4. Louter, *Windshield Wilderness*, 3; John C. Preston to G. Ballis, July 20, 1954, in Trexler, *The Tioga Road*, supra note 116.
5. Martin Litton, "Yosemite's Beauty Fast Disappearing," *National Parks Magazine* (October–December 1952): 164.
6. Litton, "Yosemite's Beauty Fast Disappearing," 168.
7. Master Plan Outline: Yosemite National Park, Introduction, 1952, Master Plans, Box 1 of 9, YNPA.
8. Stegner, *Uneasy Chair*, 297.
9. Bernard DeVoto, "National Park Service," *Harpers* (March 1949): 64–67.
10. Bernard DeVoto, "Let's Close the National Parks," *Harper's Magazine* (October 1953): 49–52.
11. Charles Stevenson, "The Shocking Truth about Our National Parks," *Reader's Digest*, January 1955; Wirth, *Parks, Politics, and the People*, 237–38; Carr, *Mission 66*, 7–10.
12. Carr, *Mission 66*, 19; Sellars, *Preserving Nature in the National Parks*, 180–83, 187–89, 201–3, 205–12; *Annual Report to the Secretary of the Interior*.
13. Greene, *Yosemite, vol.* 2, 752–55; Sellars, *Preserving Nature*, 183; Yosemite National Park, *Mission 66 Final Prospectus*, July 1956, 9–11.
14. Trexler, *Tioga Road*, 1–27.
15. Richard H. Quin, "Yosemite National Park Roads and Bridges," Historic American Engineering Record (HAER no. CA-117), 11–16; Cohen, *The History of the Sierra Club*, 9 and 95; Carr, *Mission 66*, 258–60.
16. John C. Preston to G. Ballis, July 20, 1954; Trexler, *The Tioga Road*, supra note 116.
17. American Automobile Association, *Travel News Bulletin*, no. 8, May 8, 1947; Thomas C. Vint, Chief, NPS Landscape Architecture Division, "Statement Concerning Road Development in the High Sierra Section of Yosemite National Park," Typed MSS, November 1, 1948, YNPRL.
18. Ansel Adams, "Tenaya Tragedy," *Sierra Club Bulletin* (November 1958): 1.
19. Spaulding, *Ansel Adams, and the American Landscape*, 21–22; Adams, *Ansel Adams*, 51–52.
20. Spaulding, *Ansel Adams*, 30–55; Watts, "Photography's Workshop," 123–25.
21. Szarkowski, "Introduction Ansel Adams," xii; "Realism with Reverence," *Time*, June 4, 1951, 69.
22. Taylor, *Pilgrims of the Vertical*, 54; Turner, *David Brower*, 11–12; Wyss, *The Man Who Built the Sierra Club*, 22–23.
23. Turner, *David Brower*, 11–12.

24. David Brower and Harold Bradley, "Roads in the National Parks," *Sierra Club Bulletin* 34, no. 6 (June 1949): 33–36.
25. Cohen, *The History of the Sierra Club*, 92–93; Carr, *Mission 66*, 262–67.
26. Harold Bradley, "Yosemite's Road Problem," *Pacific Discovery* (January 1951): 6–7.
27. Alexander McArthur to Chief Ranger, April 25, 1958, Central Files, Camping 1954–1976, Shelf 117, YNPA; Response to Mr. A. McArthur, April 28, 1958, 1958, Central Files, Camping 1954–1976, Shelf 117, YNPA.
28. Young, *Heading Out*, 210–213; Memorandum: Acting Superintendent, Yosemite, to Regional Director, Region Four, Trailer Camping, November 21, 1957, Central Files, Camping 1954–1976, Shelf 117, YNPA.
29. Frederick M. Martischang, Supervisory Park Ranger, United States Department of the Interior, National Park Service, "Yosemite National Park, House Trailer Use Study," June–August 1955, November 21, 1957, Central Files, Camping 1954–1976, Shelf 117, YNPA; Acting Superintendent Chas Hill to Fred Grey, Trailer Camping, November 21, 1957, Central Files, Camping 1954–1976, Shelf 117, YNPA.
30. Jan Weyl, "Time of Our Lives," *Ladies Home Journal*, July 1954, 91–120; Rugh, *Are We There Yet?*, 118–21.
31. U.S. Department of the Interior, National Park Service, in cooperation with the Californian Division of Highways State-Wide Highway Planning Survey and U.S. Department of Commerce Bureau of Public Roads, "Yosemite National Park Travel Survey," December 16, 1952–December 15, 1953, 14–17, YNPA.
32. Weyl, "Time of Our Lives," 120.
33. Weyl, "Time of Our Lives," 119.
34. Rugh, *Are We There Yet?*, 2–4.
35. Carr, *Mission 66*, 227–28; Sellars, *Preserving Nature in the National Parks*, 190–202.
36. Roper, *Camp 4*, 31–49; Taylor, *Pilgrims of the Vertical*, 125–32.
37. Jerry Gallwas, Royal Robbins, and Mike Sherrick, "Half Dome: First Ascent of the North West Face," edited by John Harlin, May 31, 2011, https://issuu.com/johnharlin/docs/gallwashalfdome9-6-10.
38. Harding, *Downward Bound*, 108.
39. O'Connell, *Beyond Risk*, 97–89; Taylor, *Pilgrims of the Vertical*, 152–58.
40. Roper, *Camp 4*, 71; William Feuerer and Warren Harding to Superintendent John C. Preston, October 3, 1957, Yosemite Central Files, Mountain Climbing, L3423-L48, YNPA.
41. Roper, *Camp 4*, 81–82.

42. E. T. Scoyen to George Whitmore, April 2, 1959, Yosemite Central Files, Mountain Climbing, L3423-L48, YNPA.
43. John C. Preston to Regional Director, November 20, 1958, Mountain Climbing, L3423-L48, YNPA.
44. Scoyen to Whitmore, April 2, 1959; John C. Preston to Regional Director, January 23, 1959, Mountain Climbing, L3423-L48, YNPA.
45. "Harding Labels Trick Climbing Charge as Ridiculous," *Merced Sun-Star*, January 27, 1959, Mountain Climbing, L3423-L48, YNPA.
46. George Whitmore to Conrad Wirth, March 16, 1959, Mountain Climbing, L3423-L48, YNPA.
47. Memorandum, Superintendent of Yosemite to Regional Director, Western Region, July 24, 1964, Yosemite Central Files, Mountain Climbing, L3423-L48, YNPA; Sax, *Mountains without Handrails*, 49; Taylor, *Pilgrims of the Vertical*, 157.
48. Elizabeth Chai Vasarhelyi, director, *Free Solo*, featuring Alex Honnold, Jimmy Chin, and Tommy Caldwell (Washington DC, 2018).
49. Stewart M. Green and Chris Van Leuven, "Profile: Ken Yager, Winner of the David R. Brower Conservation Award," *Alpinist*, April 20, 2015, http://www.alpinist.com/doc/web15s/wfeature-ken-yager-davd-brower-conservation-award; Andrew Bisharat, "Mob Mentality: Is Yosemite's El Capitan Turning into an Everest Base Camp-Size Mess?" *Outside* (May 5, 2020): 12–14.
50. Naomi Bryson, *Mountains of Memories and Myths: The Living History of the National Brotherhood of Skiers* (Xlibris LLC, 2014), "Part One Introducing the Co-Founders"; "Cocktail Chit Chat," *Jet*, February 24, 1966, 40.
51. Bryson, *Mountains of Memories and Myths*.
52. Benjamin Finley and Arthur Clay Hall of Fame Nominee, U.S. Ski and Snowboard Hall of Fame, accessed May 29, 2024, https://skihall.com/nominees/benjamin-finley-arthur-clay/.
53. Scott and Lee, "People of Color and Their Constraints to National Park Visitation."
54. Wirth, *Parks, Politics, and the People*, 359. On the Echo Park dam controversy, see: Harvey, *A Symbol of Wilderness*.
55. Ansel Adams, "Tenaya Tragedy," *Sierra Club Bulletin* (November 1958): 4.
56. Conrad Wirth, "Remarks at the Tioga Road Dedication," June 24, 1961, Addresses, United States National Park Service, University of California, Berkeley.
57. Miles, *Wilderness in the National Parks*, 137–138; Carr, "Park, Forest, and Wilderness."
58. On the history of Lake Mead as a National Recreation Area, see: Foster, *Lake Mead National Recreation Area*.

59. Childers, "Erring on the Side of Public Use," 41–65. Sax, *Mountain without Handrails*, 11; Conrad Wirth to Howard Zahnhiser, March 19, 1956, Central Files, acc 4972, attic file L3423-L48, 1955–1976, Wilderness Areas, 1955–1972, YNPA.
60. Sen. Frank Moss, "Parks Are for People," Congressional Record, U.S. Senate, August 1, 1968, 24617.
61. Sen. Frank Moss, "Parks are for People," Congressional Record, U.S. Senate, August 1, 1968, 24619.
62. Bettie Willard to Senator Frank Moss, Rocky Mountain National Park, August 15, 1968, Bettie Willard Collection, Box 3, RMNPA; Robert Barbee, "Rebuke for Chase," *Rocky Mountain News*, January 19, 1968.
63. Memorandum, Secretary Steward Udall, "Management of the National Park System," July 10, 1964, in Dilsaver, ed., *America's National Park System*, 272–76.

5. A Flood of Visitors

1. Doug Caldow, former Curry Company bellhop, interview by author, April 30, 2016.
2. "Farewell, Firefall," *Progress-Bulletin*, March 19, 1968, 1; Runte, *Yosemite*, 202.
3. Pickett quoted in "Farewell, Firefall," *Progress-Bulletin*, March 19, 1968, Sec. 2, 2.
4. Jack Goodman, "The People vs. Yosemite: Can Our Parks Be Saved?" *New York Times*, May 3, 1970.
5. George B. Hartzog Jr., "The National Parks, 1965," interview by Amelia R. Fry, April 4, 1965, Regional Oral History Office, University of California, Berkeley, 32, https://archive.org/stream/nationalparks00hartrich#page/n95/mode/2up.
6. Galen Rowell, "The Yosemite Story, On Location with MCA," *Sierra Club Bulletin* (October 1974): 39.
7. "News Release: Park Service Orders Fresh Start on Yosemite Planning Process," United States Department of the Interior, National Park Service, December 13, 1974, YNPA.
8. Statement of James O'Toole, Protection, July 4, 1970, Riot Material, Law Enforcement, General 1970, YNPA; 1970 Information on Memorial Day, July 4th, and Labor Day Weekends, 2 of 3, Law Enforcement, General 1970, YNPA; *Berkeley Tribe*, July 1970, 17–24; 1970 Information on Memorial Day, July 4th, and Labor Day Weekends, 3 of 3, YNPA.
9. Avedisian, "The Yosemite Riot," 19–20; Hummel, *Stealing the National Parks*, 235–36.
10. Statement of David Patterson, June 1, 1970, Information on Memorial Day, July 4th, Stoneman, and Labor Day Weekends, 2 of 3, Law Enforcement, General 1970, YNPA.

11. Statement of Patterson, YNPA.

12. John Fisher, MD, to President Richard Nixon, July 22, 1970, 1970 Information on Memorial Day, July 4th, and Labor Day Weekends, 1 of 3, Law Enforcement, General 1970, YNPA.

13. Avedisian, "The Yosemite Riot," 36.

14. Fisher to Nixon, July 22, 1970, YNPA.

15. Mrs. E. Ford and Family to Mr. George Hartzog Jr., 1970 Information on Memorial Day, July 4th, and Labor Day Weekends, 2 of 3, Law Enforcement, General 1970, YNPA.

16. James O'Toole, Statistics for July 2 through July 6, 1970. February 5, 1971. 1970 Information on Memorial Day, July 4th, and Labor Day Weekends, 2 of 3. Law Enforcement, General 1970. YNPA; Memorandum from Thomas Parker, Total Number of Arrests, July 4 and 5, YNPA.

17. Lawrence Hadley to Senator Alan Cranston, July 25, 1970, Law Enforcement, General 1970. YNPA.

18. Dorothy Goldeen to Director Joseph Rumburg, September 10, 1970, Law Enforcement, General 1970. YNPA.

19. Memorandum from Superintendent, Yosemite to Director, Western Region. "A Goal Oriented Approach to Law Enforcement," October 1, 1970, YNPRL; Operational Plan Labor Day Holiday period, 1970, August 17, 1970, Law Enforcement, General 1970, YNPA; "A Blade of Grass," *The Berkeley Tribe* (July 17–24, 1970): 5, YNPA; Jack Hope, "Hassles in the Park," *Natural History*, May 1971, 20–21.

20. Joseph C. Rumburg Jr. to Dorothy Goldeen, September 23, 1970, Law Enforcement, General 1970, YNPA.

21. Avedisian, "The Yosemite Riot," 57–59; Kenneth Tapman, "Investigation Report of July 4, 1970, Incident at Yosemite National Park," Washington DC: Department of Interior, 1970, Yosemite National Park Research Library, YNPRL.

22. Ladino, *Reclaiming Nostalgia*, ix; Memorandum, Director Joseph Rumburg Jr., Weapons Policy—Western Region, November 23, 1970, YNPA.

23. Sellars, *Preserving Nature in the National Parks*, 208.

24. On the Yosemite Mafia, see: Berkowitz, *Legacy of the Yosemite Mafia*; Brady, *Yosemite Mafia*; Quote from Susan P. Bratton, interview with Richard Sellars, March 20–21, 1989, in Sellars, *Preserving Nature in the National Parks*, 208, supra note 12 at 209. Bratton was a Park Service biologist in the Great Smoky Mountains National Park in the 1970s.

25. Runte, *Yosemite*, 202–4.

26. On the environmental decade, see: Rothman, *Saving the Planet*, 131–57; Steinberg, *Down to Earth*, 251–53.

27. News Release, "Yosemite Master Plan, Preliminary Draft, Rejected," August 12, 1974, 2, YNPRL; Ed Hardy to Superintendent Arnberger, June 12, 1974, Central Files, Master Plan, YNPA.
28. Hummel, *Stealing the National Parks*, 287–89.
29. "Congress Examining Yosemite Operations," *San Francisco Examiner and Chronicle*, December 8, 1974, Central Files, Concessions Box, YNPA; "Glacier Point Aerial Tramway Preliminary Feasibility Report," May 1, 1969, Central Files, Concessions Box, YNPA; Runte, *Yosemite*, 204–5.
30. Jack Anderson, "Yosemite: Another Disneyland?" *Washington Post*, September 15, 1974, B7; Hummel, *Stealing the National Parks*, 288.
31. Dilsaver and Tweed, *Challenge of the Big Trees*, 278–82 and 298–301; Sax, *Mountains without Handrails*, 67–70; Childers, *Colorado Powder Keg*, 120–21.
32. "Yosemite: Another Disneyland?" *Washington Post*, B7; Rowell, "The Yosemite Story," 18–20.
33. Arthur W. Magill, "An Overview of Campground Reservation Systems," Camping 1954–1976, YNPA.
34. Complaint from Carl Mancini, August 17, 1974, Camping and Campfire Permits, 1969–1975, YNPA.
35. James McNamara to Regional Director, National Park Service, November 14, 1974, Camping 1954–1976, YNPA.
36. Marc Janowitz to Chief Ranger, Yosemite National Park, October 23, 1974, Campground Register, 1974, YNPA.
37. James Brady to Mr. Janowitz, November 12, 1974, Campground Register, 1974, YNPA.
38. Memorandum Coyt H. Hakett, "Implementation of Test Program," November 1, 1973, Campground Register, 1974, YNPA.
39. Josephine Jackson to National Park System, August 27, 1974, Campground Register, YNPA; Luenna E. Rulison to Superintendent Leslie Arnberger, September 23, 1974, Campground Register, YNPA.
40. "News Release: Park Service Orders Fresh Start on Yosemite Planning Process," United States Department of the Interior, National Park Service, December 13, 1974, YNPA.
41. Galen Rowell, "So Long Smokey? The Yosemite Solution to Ursus Americanus," *Sierra Club Bulletin* (February 1974): 27.
42. Rowell, "So Long Smokey?" 28.
43. Charles Petit, "Yosemite Bear Killing Protested," *San Francisco Chronicle*, November 21, 1973, 4; Rowell, "So Long Smokey?" 29; Biel, *Do (Not) Feed the Bears*, 86–112.

44. Complaint, February 19, 1973, Central Files L3423-L48, YNPA.
45. Superintendent Thompson to Mr. Donald Brondyke, March 7, 1973, Central Files L3423-L48, YNPA.
46. Superintendent John Good to Mr. Ralph Johnson, December 20, 1971, Central Files L3423-L48, YNPA.
47. Carlos A. Camargo, MD, to Superintendent, Yosemite National Park, March 12, 1971, Central Files L3423-L48, YNPA; *Tehipite Topics* 17, no. 3, March 1970.
48. U.S. Department of the Interior, National Park Service, "Snowmobile Ban at Yosemite to Continue This Winter," News Release, February 29, 1975, Central Files L3423-L48, YNPA.
49. Sierra Snowmobile Club to Leslie P. Arnberger, January 15, 1975, Central Files L3423-L48, YNPA; Leslie P. Arnberger to Don Headrick, February 11, 1975, Central Files L3423-L48, YNPA.
50. James P. McCarthy to Mr. Harthon Bill, Director, National Park Service, June 24, 1971; Memorandum, to Superintendent through Valley District Ranger and Valley District Manager, "Meeting with R. Bedayn, American Alpine Club," July 27, 1971, YNPA.
51. Letter to the Editor, *San Francisco Chronicle*, December 1, 1970; E. T. Scoyen to Edward Hummel, March 4, 1971, Central Files L3423-L48, YNPA.
52. Lawrence Hadley to Jeff Foott, February 12, 1969, YNPA; Superintendent John Preston to Mike Harding, November 25, 1964, Central Files L3423-L48, YNPA.
53. Roper, *Camp 4*, 212; Taylor, *Pilgrims of the Vertical*, 244.
54. Kirk, *Counterculture Green*, 198–203; Gross, *Shopping All the Way to the Woods*, 165–166.
55. Johnson, "Wilderness Handrails"; Pete Thompson, "Tragedies in Yosemite Valley," *Summit* 19 (June 1973): 1–2; Pete Thompson, Assistant Valley District Ranger, to All Valley Personnel, re: "Rescue Site Guidelines," September 27, 1971, Yosemite Search and Rescue (YOSAR) office files.
56. Pete Thompson and Ralph A. Robinson, "Search and Rescue Incident Report," July 16, 1973, YOSAR files; John Long, Letter to the Editor in *Summit*, December 1973, 32.
57. James Brady to Sunnyside "Camp 4" users, re: "Management of Sunnyside Campground," February 25, 1974, YOSAR office files; James Brady, "Open Letter to Yosemite Climbers," *Summit*, March 1974, 1, YOSAR office files.
58. Spence, *Dispossessing the Wilderness*, 130–31.
59. Cobb, *Native Activism in Cold War America*, 1–7; Blansett, *A Journey to Freedom*, 1–10; Deloria Jr. *Custer Died for Your Sins*.
60. Memorandum to Files from Leslie Arnberger, Superintendent, Yosemite National Park, September 29, 1976, YNPRL.

61. Summary of Meeting with American Indian Council of Mariposa County, Files (L3035 X–K1815), December 21, 1976, Superintendent Leslie Arnberger, YNPRL.
62. Memorandum P4019 to Personnel Officer from Superintendent Leslie P. Arnberger, December 20, 1976, YNPRL; Notes about the Indian Cultural Program Supervisor from the Superintendent, January 31, 1977, YNPRL.
63 Proposal to the National Park Service, *The Indian Perspective on the Future of Yosemite National Park*, submitted by American Indian Council of Mariposa County, revised version, February 1978, YNPRL.
64. Barbara Karshmer to Superintendent Les Arnberger, July 31, 1978, YNPA.
65. Leslie P. Arnberger to Barbara Karshmer, August 10, 1978, Central Files, acc. 5586, YNPA.
66. Memo to John Bryne, Assistant Superintendent, from Craig D. Bates, Indian Cultural Program Specialist, (K1815) November 9, 1978, Central Files, acc. 5586, YNPA; Memorandum to All Employees from Leslie Arnberger, Superintendent, May 29, 1980, Central Files, acc. 5586, YNPA.
67. "Park Service Orders Fresh Start on Yosemite Planning Process," News Release, December 13, 1974, YNPRL.
68. Leach Nichols, "Many Ideas Offered at Yosemite Planning Meeting," *Bridgeport Chronicle Union*, February 20, 1975; "Yosemite Master Plan Mulled," *Argus Courier*, March 31, 1975; Ken Schultz, "County Folks Speak Out on Yosemite Plans," *Californian*, April 10, 1975; "Yosemite Suggestions Made," *Long Beach Press Telegram*, April 23, 1975.
69. Robert A. Jones, Vision of Yosemite's Future Emerges, Public Attitudes toward Park Development Analyzed," *Los Angeles Times*, July 20, 1976, B3 & B17

6. Absolutely Democratic

1. James M. Dodds to Chief Ranger, Yosemite National Park, August 17, 1977, Central Files, acc. 5586, YNPA.
2. Mrs. Beulah E. Bennet to Mr. Cecil Andrus, Secretary of the Interior, August 24, 1977, Central Files, acc. 5586, YNPA; Dr. David Kalbfell to Congressman Bob Wilson, August 5, 1977, YNPA; Donald Jones to Superintendent Leslie Arnberger, October 8, 1978, Central Files, acc. 5586, YNPA; Memorandum to Superintendent, Assistant Superintendents, and Division Chiefs from Landscape Architect, April 3, 1979, Central Files, acc. 5586, YNPA.
3. "NPS Orders Fresh Start on Yosemite Planning Process," News Release, December 13, 1974, Central Files, acc. 5586, YNPA.
4. Allen Wicker, "Crowding on Yosemite Buses," *Science*, July 11, 1975, 88.

5. Harry Coulter, "Yosemite: Beautiful Land of Uncertain Tomorrows," *Auto Club News Pictorial*, August 1977, 8.
6. United States Department of the Interior, National Park Service, "Yosemite General Management Plan," September 16, 1980, 1, YNPA.
7. Howard H. Chapman, Memo on October 5, 1978, Central Files, acc. 5586, Y14-Y46, YNPA; Department of Fish and Game, April 21, 1982, Central Files, acc. 5586, Y14-Y46, YNPA.
8. Bruce Porter, "I'm Not Coming Down!" *New York Times*, August 15, 1974, 169; Ronald Taylor, "National Park Policy," *New York Times*, November 7, 1975, 37.
9. Yosemite Hang Gliding Association, accessed May 4, 2018, www.yhga.org.
10. Ralph Suelzle to Superintendent Robert Binnewies, January 25, 1980, Central Files D18, 1977–1984, YNPA; Superintendent Robert Binnewies to Ralph Suelzle, February 11, 1980, Central Files D18, 1977–84, YNPA.
11. Yosemite Hang Gliding Association, www.yhga.org.
12. Draft General Management Plan, Department of the Interior, Washington DC, 4. YNPA.
13. William Moffitt Chairman, Mariposa County Board of Supervisors to Chairman Jack Brooks, April 14, 1976, YNPA.
14. Madalene Good, President, Mono County Chamber of Commerce, to United States Department of Interior, February 13, 1976, YNPA.
15. Helvarg, *The War against the Greens*, 147.
16. Statement of George B Hartzog Jr., Director, National Park Service, Department of the Interior, before the Subcommittee on Parks and Recreation of the House Interior and Insular Affairs Committee, January 18, 1968, YNPA; Spence, *Watershed Park*, 195–99; Watt, *The Paradox of Preservation*, 187–213.
17. Berkowitz, *Legacy of the Yosemite Mafia*, 147–48.
18. Helvarg, *The War against the Greens*, 147; Berkowitz, *Legacy of the Yosemite Mafia*, 140–151.
19. Milt Rowley to The Honorable Alan Cranston, June 9, 1975, YNPA; John H. Davis, Acting Regional Director, Western Region, to Honorable Thomas Bees, August 6, 1974, YNPA.
20. Mariposa County Board of Supervisors to Ronald H. Walker, August 26, 1974, YNPA.
21. "Policy Shift Angers U.S. Park Residents," *Los Angeles Times*, June 18, 1978; Richard Cockle, "Cattle-Grazing Advocates Picket Environmentalists' Conference," *The Oregonian*, April 27, 1992; "Land Grab by the Parks," *Newsweek*, August 14, 1978, 21; Robert Jones, "Policy Shift Angers U.S. Park Residents," *Los Angeles*

Times, June 18, 1978; Switzer, *Green Backlash*, 11–12; "Two Challenge Park Land Purchase Plan," *Akron Beacon Journal*, July 19, 1978.

22. "Two Challenge Park Land Purchase Plan," *Akron Beacon Journal*, July 19, 1978; Chuck Williams, "The Park Rebellion: Charles Cushman, James Watt, and the Attack on the National Parks," *Not Man Apart*, June 1982, 11–26.
23. Hartzog, *Battling for the National Parks*, 5; Switzer, *Green Backlash*, 11–12; Skillen, *This Land is My Land*, 45–84.
24. James Watt, "Keynote Address to Outdoor Writers Association of America," *Newsweek*, June 29, 1981, Wilderness Society Series, 12, Box 3, FF29, Watt, James G. Sec of Interior 1981, Denver Public Library Conservation Collection; Childers, "Erring on the Side of Public Use," 41–65.
25. Peter Steinhart, "Interior Motives: Will Watt Get His Way in the Parks?" *Los Angeles Times*, May 3, 1981, G2.
26. Wallace Stegner, "The Best Idea We Ever Had," *Wilderness* 46 (Spring, 1983): 7–8. Five years prior Yosemite Superintendent Leslie Arnberger explained the park's policy to the *Los Angeles Times*, stating, "The Park Service simply does not believe that private towns should exist inside a national park." "Policy Shift Angers U.S. Park Residents," *Los Angeles Times*, June 18, 1978; Childers, "Returning to the Best Idea We Ever Had," 109–132.
27. Binneweis, *Your Yosemite*, 256–58; Berkowitz, *Legacy of the Yosemite Mafia*, 127–29.
28. Spaulding, "Yosemite and Ansel Adams," 615; Quotes in Robert Turnage, "Ansel Adams: The Role of the Artist in the Environmental Movement," *The Living Wilderness*, March 1980, 8 and 13.
29. Quoted in *Yosemite: The Fate of Heaven*, directed by Jon Else, produced by the Jonathan Company and the Sundance Institute in association with WGBH and WNET, Public Television, 1988.
30. National Park Service, *A Sense of Place*, 41.
31. "Fires Ravage Yosemite Lightning Strikes Trigger Blazes," *The Sun Sentinel*, August 11, 1990, 3A; Rothman, *A Blazing Heritage*, 185.
32. Smith, *Media and Apocalypse*, 39.
33. Rothman, *A Blazing Heritage*, 187–205; Pyne, *Pyrocene Park*, 129–50.
34. Katherine Bishop, "Closing of Yosemite Renews Debate over Fire Policies," *New York Times*, August 16, 1990, A18.
35. Bramwell, *Wilderburbs*, 117–176.
36. Jack Goldsmith, "Designing for Diversity," *National Parks* (May–June 1994): 20–21; Finney, *Black Faces, White Spaces*, 104. For African American experiences within the National Parks, see: Young, *Heading Out*, 173–207.

37. Guy Lucier, Letter to the Editor, *National Parks* 68 (September–October 1994): 6; Joseph Lange, Letter to the Editor, *National Parks* 68 (September–October 1994): 6.
38. Mireya Navarro, "National Parks Reach Out to Blacks Who Aren't Visiting," *New York Times*, November 2, 2010, A17.
39. Ken Burns, director, *The National Parks: America's Best Idea, Episode One* (Arlington: PBS Home Video, 2009).
40. Glenn Nelson, "How Shelton Johnson Became the Buffalo Soldiers' Champion," *High Country News*, July 28, 2016.
41. Navarro, "National Parks Reach Out to Blacks Who Aren't Visiting."
42. "When Oprah Met Shelton," Oprah Winfrey Network, October 29, 2010, https://www.oprah.com/own-oprahshow/oprah-and-gayle-arrive-at-yosemite-national-park-video.
43. "Ranger Shelton Looks Back at Oprah and Gayle's Camping Adventure, Where Are They Now?" OWN, accessed September 16, 2024, https://www.youtube.com/watch?v=ajGom5w_JuY.
44. Stan Wacht, quoted in Goodin, *Yosemite*.
45. Michael McCabe, "Yosemite Closed Indefinitely," *San Francisco Chronicle*, January 9, 1997.
46. Stoll, *Inherent the Holy Mountain*, 99; Anderson, *The Nature of Church Camp*, 29–32.
47. Rose Certini, "Yosemite's Splendor Backdrop for Worship," *Merced Sun Star*, July 28, 1990, 14; "Popularity Worries Yosemite Rangers," *Merced Sun Star*, July 28, 1990, 15.
48. "Popularity Worries Yosemite Rangers," *Merced Sun Star*, July 28, 1990, 15.
49. "Superintendent's Annual Report," Yosemite National Park (1992), 17, YNPRL.
50. U.S. Department of the Interior, National Park Service, Sentinel Bridge, Yosemite National Park, California, *Environmental Assessment to Replace Sentinel Bridge*, July 1989, 370, YNPA.
51. Wild and Scenic River Act, 1968 PL 90–542 (1 USC 1271–1287).
52. Tim Golden, "In Yosemite, Nature May Have Its Way Yet," *New York Times*, February 2, 1997, 1.
53. Carl Nolte, "Babbitt Tells Details of Yosemite Changes/Less Park, Lodging—More Shuttle Buses," *San Francisco Chronicle*, March 28, 2000, A9.
54. "Brower on the Yosemite Valley Plan," *San Francisco Chronicle*, November 20, 2000, A19.
55. Kirk and Palmer, "When Nature Becomes Culture," 505.
56. United States Department of the Interior, National Park Service, Yosemite National Park, "Summary of Public Comments and Responses Yosemite Val-

ley Plan/SEIS, Yosemite National Park," in *Yosemite Valley Plan Supplemental Environmental Impact Statement*, vol 3.

57. Brian Skoloff, "Yosemite 'Remodeling Project' Controversial," *Los Angeles Times*, December 21, 2003.
58. Glen Martin, "Yosemite Once Again a Battleground: Mariposa Congressman Wants to Change Clinton-Era Plan," *San Francisco Chronicle*, July 27, 2003, A19.
59. Manning, *Parks and Carrying Capacity*, 110–24; Cathcart-Rake, "The Friends of Yosemite Valley Saga."
60. Whitney, *The Yosemite Book*, 67; Kirk and Bramwell, *The Yosemite Way*, 389–90.
61. National Park Service, U.S. Department of the Interior, Yosemite National Park, "Half Dome Trail Stewardship Plan Environmental Assessment," January 2012, 1–4.
62. Peter Fimrite, "Danger on the Dome, Overcrowding: Hikers Swarming Yosemite's Half Dome Create a Bottleneck at the Treacherously Steep Granite Climb to the Summit," *San Francisco Chronicle*, July 7, 2007.
63. Peter Fimrite, "Deadly Trek up Half Dome / Rangers Re-examining Safety of Popular Hike a Fatal Fall from Cables During Final Ascent," *San Francisco Chronicle*, June 19, 2007.
64. "Half Dome Trail Stewardship Plan Environmental Assessment," January 2012, 1–10; Half Dome Environmental Assessment Redacted Public Comments, https://home.nps.gov/yose/getinvolved/upload/HalfDomeeacomments-Redacted_Redacted.pdf.
65. National Park Service, U.S. Department of the Interior, Yosemite National Park, "Half Dome Trail Stewardship Plan: Finding of No Significant Impact," December 2012, 2–3.
66. Wilderness Act of 1964, Public Law 88–577, 88th Congress, 2nd sess. (September 3, 1964), 2.
67. Julie Cart, "Park Rangers over the Line at Yosemite? Critics See Pattern of Abusing their Authority as Police," *Los Angeles Times*, December 26, 2004.
68. John Lancaster, "Trouble in Paradise: National Park Rangers Growing Burden of Police Work," *Washington Post*, June 24, 1991; "Motel Worker Charged with Yosemite Killing: Cary Stayer is Also a Suspect in the Murder of Three Female Tourists in the National Park Last February," *Vancouver Sun*, July 27, 1999, A5; Laura Sullivan, "The Dangers of Being a Ranger," National Public Radio, June 18, 2005, http://www.npr.org/templates/story/story.php?storyId=4709159.
69. "One Week Time Limit," SuperTopo, accessed November 27, 2018, http://www.supertopo.com/climbers-forum/39517/One-Week-Time-Limit-in-the-Valley.
70. John Branch, "Lost Brother in Yosemite," *New York Times*, June 11, 2015, SP 1.

71. Terra Galleri Photography, accessed May 19, 2018, www.terragalleria.com/mountain/info/yosemite/base.html.
72. "Official Statement from Patagonia and Dean Potter on the Delicate Arch Climb," *Outside*, June 8, 2006, accessed September 23, 2024, https://www.outsideonline.com/outdoor-adventure/climbing/official-statement-patagonia-and-dean-potter-delicate-arch-climb; "Dean Potter's GoPro Captured Fatal Wingsuit Jump," *Outside*, May 20, 2015, accessed September 23, 2024, https://www.outsideonline.com/outdoor-adventure/dean-potters-gopro-captured-fatal-wingsuit-jump/.
73. Charlotte Simmonds, Annette McGivney, Patrick Reilly, et al., "Crisis in Our National Parks: How Tourists Are Loving Nature to Death," *The Guardian*, November 20, 2018; Erin McCormick and Michael Safi, "In Yosemite, Two Deaths Raise Concerns about Social Media," *High Country News*, November 6, 2018.
74. Wilkins, Smith, and Keane, "Social Media Communication Preferences," 4–18.
75. The United States of the Interior, National Park Service, Yosemite National Park, "Restoration of the Mariposa Grove of Giant Sequoias Final Environmental Statement," October 2013; Paul Rogers, "Yosemite Grove Rehab a Nod to the Park's Past," *Chicago Tribune*, June 2014, 8.
76. The United States Department of the Interior, National Park Service, Yosemite National Park, "Comment Letters Received during the 60-day Public Review of the Draft Environmental Impact Statement or the Restoration of the Mariposa Grove of Giant Sequoias Environmental Statement," May 6 and 7, 2013.
77. Yosemite National Park, News Release, June 14, 2018, National Park Service, https://www.nps.gov/yose/learn/news/yosemite-national-park-and-yosemite-conservancy-celebrate-the-reopening-of-the-mariposa-grove-of-giant-sequoias-after-the-largest-restoration-project-in-park-history.htm.

Epilogue

1. "Yosemite National Park Continued a Phased Reopening on Thursday, June 11, 2020, News Release, June 8, 2020, National Park Service, https://www.nps.gov/yose/learn/news/yosemite-national-park-continues-a-phased-reopening-on-thursday-june-11-2020.htm.
2. Kurtis Alexander, "Yosemite's Grand Opening Thrills Flocks of Visitors," *San Francisco Chronicle*, June 12, 2020, A1.
3. Carmen George, "Day-use Reservations Needed at Yosemite," *Fresno Bee*, April 9, 2021, 5.
4. Interview with Emily Dayhoff, November 24, 2021.
5. Jim Carlton, "Yosemite Might Bring Back Reservations as Park Is Overwhelmed with Crowds Again," *Wall Street Journal*, July 23, 2023, https://www.wsj.com

/articles/yosemite-might-bring-back-reservations-as-park-is-overwhelmed-with-crowds-again-98e759b4.

6. Michael Brune, "Pulling Down Our Monuments," Sierra Club, July 22, 2020, https://www.sierraclub.org/michael-brune/2020/07/john-muir-early-history-sierra-club; Fleck, "John Muir's Evolving Attitudes," 21.
7. Julian Brave Noise Cat, "The Environmental Movement Needs to Reckon with its Racist History," Vice, September 13, 2019, https://www.vice.com/en/article/bjwvn8/the-environmental-movement-needs-to-reckon-with-its-racist-history.
8. John Muir, *Our National Parks*, 1; John Muir to Sarah. Muir Galloway, September 3, 1873, University of the Pacific Library, Holt-Atherton Special Collections.
9. Spence, *Dispossessing the Wilderness*, 115–132; Cothran, "Working the Indian Field Days"; Burnham, *Indian Country, God's Country*, 19–21.
10. Mendoza, "Review of *City of Inmates*," 399.
11. Hardin, "The Tragedy of the Commons," 1245.
12. For opposing views, see: Tweed, *Uncertain Path*; Soukup and Machlis, *American Covenant*.
13. John Garder and Angela Gonzales, "How the New Administration's Actions Will Affect National Parks," National Parks Conservation Association Blog, February 4, 2025, https://www.npca.org/articles/6680-how-the-new-administration-s-actions-will-affect-national-parks.
14. Emily Yehle, "Park Service Grapples with "Frustrated" Workforce," *Politico*, May 31, 2016, https://www.eenews.net/articles/park-service-grapples-with-frustrated-workforce/.
15. Gavin Feek, "Why Did Rangers Fly an Upside-Down American Flag from Yosemite's Horsetail Fall?" *Outside*, February 24, 2025, https://www.outsideonline.com/outdoor-adventure/climbing/yosemite-climbing-layoffs.
16. Wallace Stegner, "The Best Idea We Ever Had: An Overview," *Wilderness* 46 (Spring 1983): 4.

Bibliography

Archives

Francis Loeb Library, Harvard University, Cambridge MA

Papers of John Olmsted Collection

Huntington Library, San Marino CA

William J. Howard Papers

National Archives, Washington DC

Record Group 17 (Yosemite)

University of the Pacific Library Holt-Atherton Special Collections, Stockton CA

John Muir Papers

Yosemite National Park Archives YNPA

Calvary Administration Collections

Central Files Collection

Law Enforcement Collection

Old Central Files Collection

Resource Management Papers Collections

Yosemite National Park Research Library YNPRL

Yosemite Search and Rescue Archive YOSAR

Published Sources

Adams, Ansel. *Ansel Adams: An Autobiography*. New York: Little Brown and Company, 1985.

Akins, Damon, and William Bauer Jr. *We Are the Land: A History of Native California*. Berkeley: University of California Press, 2021.

Albright, Horace M., and Robert Cahn. *The Birth of the National Park Service: The Founding Years, 1913–33*. Vol. 2 of Institute of the American West Books. Salt Lake City: Howe Bros., 1985.

Albright, Horace M., Marian Albright Schenck, and Robert M. Utley. *Creating the National Park Service: The Missing Years*. Norman: University of Oklahoma Press, 1999.

Anderson, Christopher. *The Nature of Church Camp: An Environmental History of Outdoor Ministry, 1945–1980*. Lanham: Lexington Books, 2023.

Annual Report to the Secretary of the Interior for the Fiscal Year Ending June 30, 1956. Washington DC: Government Printing Office, 1956.

Avedisian, Laura. "The Yosemite Riot: Changes in Policy and Management in the National Park Service." Master's thesis, San Jose State, 1998.

Badè, William Fredric. *The Life and Letters of John Muir*. Boston: Houghton Mifflin Company, 1924.

Bates, Craig. "Native Americans Perpetuating a Wealth of Cultures in the Parks." *National Parks* 59 (September/October 1985): 12–15.

Bates, Craig, and Martha Lee. *Tradition and Innovation: A Basket History of the Indians of the Yosemite–Mono Lake Area*. Yosemite Association, 1991.

Beckert, Sven. "Propertied of a Different Kind: Bourgeoise and Lower Middle Class in the Nineteenth-Century United States." In *The Middling Sorts: Explorations in the History of the American Middle Class*, edited by Burton Blendstein and Robert Johnston. New York: Routledge, 2001.

Berkowitz, Paul. *Legacy of the Yosemite Mafia: The Ranger Image and Noble Cause Corruption in the National Park Service*. Walterville OR: Trine Day LLC, 2017.

Biel, Alice Wondrak. *Do (Not) Feed the Bears: The Fitful History of Wildlife and Tourists in Yellowstone*. Lawrence: University Press of Kansas, 2006.

Binnewies, Robert O. *Your Yosemite: A Threatened Public Treasure*. Ashland OR: White Cloud Press, 2015.

Blansett, Kent. *A Journey to Freedom: Richard Oakes, Alcatraz, and the Red Power Movement*. New Haven: Yale University Press, 2018.

Bledstein, Burton. "Introduction: Storytellers to the Middle Class." In *The Middling Sorts: Exploration in the History of the American Middle Class*, edited by Burton Bledstein and Robert Johnston. New York: Routledge, 2001.

Bledstein, Burton J., and Robert D. Johnston. *The Middling Sorts: Explorations in the History of the American Middle Class*. New York: Routledge, 2001.

Blodgett, Peter, ed. *Motoring West*. Vol. 1, *Automotive Pioneers, 1900–1909*. Norman: University Press of Oklahoma, 2015.

Blodgett, Peter. "Visiting the 'Realm of Wonder': Yosemite and the Business of Tourism, 1855–1916." In *Yosemite and Sequoia: A Century of California National Parks*, edited by Richard Orsi and Alfred Runte. Berkeley: University of California Press, 1993.

Brady, James. *Yosemite Mafia: The Turbulent 1970s Ranger Challenges and Park Changes*. Published by the author, 2017.

Bramwell, Lincoln. *Wilderburbs: Communities on Nature's Edge*. Seattle: University of Washington Press, 2016.

Brechin, Gray. *Imperial San Francisco: Urban Power, Earthly Ruin*. Berkeley: University of California Press, 1999.

Browning, Peter. *Yosemite Place Names: The Historic Background of Geographic Names in Yosemite National Park*. Lafayette CA: Great West Books, 1988.

Bryson, Naomi. *Mountains of Memories and Myths: The Living History of the National Brotherhood of Skiers*. Xlibris LLC, 2014.

Bunnell, Lafayette Houghton. *Discovery of the Yosemite, and the Indian War of 1851, Which Led to That Event*, 3rd ed. New York: Fleming H. Revell Company, 1892.

Burgess, Jack. *Trains to Yosemite*. Wilton CA: Signature Press, 2004.

Burnham, Philip. *Indian Country, God's Country: Native Americans and the National Parks*. Washington DC: Island Press, 2000.

Cain, Robert. *The Birth of the National Park Service: The Founding Years, 1913–33*. Salt Lake City: The Howe Brothers, 1985.

Campbell, Colin. *The Romantic Ethic and the Spirit of Modern Consumerism*. London: Alcuin Academics, 2005.

Carr, Ethan. *Mission 66: Modernism and the National Park Dilemma*. Amherst: University of Massachusetts Press, 2007.

———. "Park, Forest, and Wilderness." *George Wright Forum* 17, no. 2 (2000): 22.

———. *Wilderness by Design: Landscape Architecture and the National Park Service*. Lincoln: University of Nebraska Press, 1999.

Carstensen, Vernon. *The Public Lands: Studies in the History of the Public Domain*. Madison: University of Wisconsin Press, 1963.

Cathcart-Rake, John. "The Friends of Yosemite Valley Saga: The Challenges of Addressing the Merced River's User Capacities." *Environmental Law* 39, no. 3 (Summer 2009): 833–66.

Chan, Yenyen F. "Interpreting the Contributions of Chinese Immigrants in Yosemite National Park's History." *The George Wright Forum* 34, no. 3 (2017): 299–307.

Childers, Michael. *Colorado Powder Keg: Ski Resorts and the Environmental Movement*. Lawrence: University Press of Kansas, 2012.

———. "Erring on the Side of Public Use: James Watt and the National Parks." *Montana: The Magazine of Western History* 73, no. 2 (Summer 2023): 41–65.

———. "For Public Use, Resort, and Recreation: The Struggle Over Appropriate Recreation in Yosemite," *George Wright Forum* 35, no. 2 (November 2018): 304–11.

———. "Returning to the Best Idea We Ever Had." In *Wallace Stegner's Unsettled Country: Ruin, Realism, and Possibility in the American West*, edited by Mark Fiege, Michael Lansing, and Leisl Carr Childers. Lincoln: University of Nebraska Press, 2024.

Chung, Sue Fawn. *Chinese in the Woods: Logging and Lumbering in the American West*. Urbana: University of Illinois Press, 2015.

Cobb, Daniel. *Native Activism in Cold War America: The Struggle for Sovereignty*. Lawrence: University Press of Kansas, 2008.

Cohen, Michael P. *The History of the Sierra Club, 1892–1970*. San Francisco: Sierra Club Books, 1988.

Cohen, Michael P., and Valerie Mendenhall Cohen. *Granite and Grace: Seeking the Heart of Yosemite*. Reno: University of Nevada Press, 2019.

Colton, Craig, and Lary Dilsaver. "The Hidden Landscape of Yosemite National Park." *Journal of Cultural Geography* 22, no. 2 (Spring/Summer 2005): 27–50.

Cothran, Boyd. "Working Indian Field Days: The Economy of Authenticity and the Question of Agency in Yosemite Valley." *American Indian Quarterly* 34, no. 2 (Spring 2010): 194–223.

Cronon, William. "A Place for Stories: Nature, History, and Narrative." *Journal of American History* 78, no. 4 (March 1992): 1347–76.

Delora, Phil. *Playing Indian*. New Haven: Yale University Press, 1999.

Deloria Jr., Vine. *Custer Died for Your Sins: An American Indian Manifesto*. New York: MacMillan, 1969.

Demars, Stanford E. *The Tourist in Yosemite, 1855–1985*. Salt Lake City: University of Utah Press, 1991.

Deverell, William. "'Niagara Magnified': Finding Emerson, Muir, and Adams in Yosemite." In *Yosemite: Art of an American Icon*, edited by Amy Scott. Berkeley: University of California Press, 2006.

Diamant, Rolf, and Ethan Carr. *Olmsted and Yosemite: Civil War, Abolition, and the National Park Idea*. Amherst: Library of American Landscape History, 2022.

Dilsaver, Lary, ed. *America's National Park System: The Critical Documents*. London: Rowman & Littlefield Publishers, 1994.

Dilsaver, Lary, and William Tweed. *Challenge of the Big Trees: A Resource History of Sequoia and Kings Canyon National Parks*. Three Rivers CA: Sequoia Natural History Association, Inc., 1990.

Drabelle, Dennis. *The Power of Scenery: Federick Law Olmsted and the Origin of National Parks*. Lincoln: University of Nebraska Press, 2021.

Eccleston, Robert, and C. Gregory Crampton. *The Mariposa Indian War, 1850–1851: Diaries of Robert Eccleston: The California Gold Rush, Yosemite, and the High Sierra*. Bancroft Library Publications, no. 6. Salt Lake City: University of Utah Press, 1957.

Finney, Caroline. *Black Faces, White Spaces: Reimagining the Relationship of African Americans to the Great Outdoors*. Chapel Hill: University of North Carolina Press, 2014.

Fleck, Richard. "John Muir's Evolving Attitudes Toward Native American Cultures." *American Indian Quarterly* 4, no. 1 (February 1978): 19–31.

Foster, Jonathan. *Lake Mead National Recreation Area: A History of America's First National Playground*. Reno: University of Nevada Press, 2016.

Frisch, Michael. *A Shared Authority: Essays on the Craft and Meaning of Oral and Public History*. New York: SUNY Press, 1990.

Goetzmann, William H., and William N. Goetzmann. *The West of the Imagination*. New York: W.W. Norton & Company, 1986.

Goodin, Mark. *Yosemite: The 100 Year Flood: Movement in Tides*. Mariposa CA: Sierra Press, 1997.

Greeley, Horace. *An Overland Journey from New York to San Francisco in the Summer of 1859*. Lincoln: University of Nebraska Press, 1999.

Green, Tyler. *Carleton Watkins: Making the West American*. Oakland: University of California Press, 2018.

———. *Emerson's Nature and the Artists: Idea as Landscape, Landscape as Idea*. Munich: Prestel, 2021.

Greene, Linda. *Yosemite Historic Resource Study: A History of the Discovery, Management, and Physical Development of Yosemite National Park*. 3 vols. Washington DC: U.S. Department of the Interior, National Park Service, 1987.

Gross, Rachel. *Shopping All the Way to the Woods: How the Outdoor Industry Sold Nature to America*. New Haven: Yale University Press, 2024.

Hampton, H. Duane. *How the U.S. Cavalry Saved the National Parks*. Bloomington: Indiana University Press, 1971.

Hardin, Garrett. "The Tragedy of the Commons." *Science* 162, no. 3859 (December 13, 1968): 1243–48.

Harding, Warren. *Downward Bound: A Mad! Guide to Rock Climbing*. Upper Saddle River NJ: Menasha Ridge Press, 1976.

Hartzog Jr., George B. *Battling for the National Parks*. Mt. Kisco NY: Moyer Bell Unlimited, 1988.

Harvey, Mark. *A Symbol of Wilderness: Echo Park and the America Conservation Movement*. Seattle: University of Washington Press, 2000.

Hays, Samuel P. *Conservation and the Gospel of Efficiency*. Pittsburgh: University of Pittsburgh Press, 1999.

Helvarg, David. *The War against the Greens: The "Wise Use" Movement, the New Right, and Anti-Environmental Violence*. San Francisco: Sierra Club Books, 1994.

History of the United States Naval Special Hospital. Yosemite National Park: Yosemite Park and Curry Company, 1946.

Hummel, Don. *Stealing the National Parks: The Destruction of Concessions and Park Access*. Bellevue WA: Free Enterprise Battle Book, 1987.

Huntley, Jen A. *The Making of Yosemite: James Mason Hutchings and the Origin of America's Most Popular Park*. Lawrence: University Press of Kansas, 2011.

Huth, Hans. *Yosemite, the Story of an Idea: An Account of the Changes in the American Attitude toward Nature That Led to the Creation of the National Park System*. Yosemite National Park: Yosemite Natural History Association, 1984, 1948.

Hyde, Anne F. "From Stagecoach to Packard Twin Six: Yosemite and the Changing Face of Tourism, 1880–1930." *California History* 69, no. 2 (Summer 1990): 154–69.

Ise, John. *Our National Park Policy: A Critical History*. Baltimore: Johns Hopkins University Press, 1961.

———. *United States Oil Policy*. New Haven: Yale University Press, 1926.

Jackson, Helen Hunt. *Ah-Wah-Ne Days: A Visit to the Yosemite Valley in 1872*. San Francisco: The Book Club of California, 1971.

Jacoby, Karl. *Crimes against Nature: Squatters, Poachers, Thieves, and the Hidden History of American Conservation*. Berkeley: University of California Press, 2001.

Jenkins, Henry. *Convergence Culture: Where Old and New Media Collide*. New York: New York University Press, 2006.

Jenkins, Jeffrey. "Science and the Evolving Management of Environmental Hazards at Yosemite National Parks." *Parks and Stewardship Forum* 38, no. 3 (September 15, 2022): 540–55.

Johnson, Christopher. "Wilderness Handrails: The Evolution of Search and Rescue in Yosemite National Park." Master's thesis, University of Nevada, Las Vegas, 2007.

Johnston, Hank. *Shortline to Paradise: The Story of the Yosemite Valley Railroad*. Long Beach CA: Johnston & Howe, 1962.

———. *Yosemite's Yesterdays*. Yosemite CA: Flying Spur Press, 1989.

———. *Yosemite's Yesterdays*. Volume 2. Yosemite CA: Flying Spur Press, 1991.

Jones, Craig H. *The Mountains That Remade America: How Sierra Nevada Geology Impacts Modern Life*. Oakland: University of California Press, 2017.

Jones, Holway R. *John Muir and the Sierra Club: The Battle for Yosemite*. San Francisco: Sierra Club Books, 1965.

Keiter, Robert B. *To Conserve Unimpaired: The Evolution of the National Park Idea*. Washington DC: Island Press, 2013.

Kelly, Tara Kathleen. *The Hunter Elite: Manly Sport, Hunting Narratives, and American Conservation*. Lawerence: University Press of Kansas, 2018.

Kiester Jr., Edwin. *Donald B. Tresidder: Sanford's Treasure* Stanford CA: Stanford Historical Society, 1992.

King, Thomas Starr. *A Vacation among the Sierras: Yosemite in 1860*. Edited by John A. Hussey. San Francisco: The Book Club of California, 1962.

Kirk, Andrew. *Counterculture Green: The Whole Earth Catalog and American Environmentalism*. Lawrence: University Press of Kansas, 2007.

Kirk, Andrew, and Lincoln Bramwell. *The Yosemite Way: An Administrative History of Yosemite National Park*. National Park Service, U.S. Department of the Interior, History Program, Pacific West Region, 2009. Unpublished.

Kirk, Andrew, and Charles Palmer. "When Nature Becomes Culture: The National Register and Yosemite's Camp 4, A Case Study." *Western Historical Quarterly* 37, no. 4 (Winter 2006): 496–506.

Kurutz, Gary. "Yosemite on Glass." In *Yosemite: Art of an American Icon*. Edited by Amy Scott, Brian Bibby, William Deverell, Kate Nearpass Ogden, Jonathan Spaulding, and Jennifer Watts. Berkeley: University of California Press, 2006.

Ladino, Jennifer. *Reclaiming Nostalgia: Longing for Nature in American Literature*. Charlottesville VA: University of Virginia Press, 2012.

Louter, David. *Windshield Wilderness: Cars, Roads, and Nature in Washington's National Parks*. Seattle: University of Washington Press, 2006.

Madley, Benjamin. *An American Genocide: The United States and the California Indian Catastrophe*. New Haven: Yale University Press, 2016.

Maher, Neil. *Nature's New Deal: The Civilian Conservation Corps and the Roots of the American Environmental Movement*. New York: Oxford University Press, 2008.

Manning, Robert. *Parks and Carrying Capacity: Commons Without Tragedy*. Washington DC: Island Press, 2007.

Matthews, Glenna. *The Golden State in the Civil War: Thomas Starr King, the Republican Party, and the Birth of Modern California*. New York: Cambridge University Press, 2012.

McClelland, Linda Flint. *Building the National Parks: Landscape and Construction*. Baltimore: Johns Hopkins University Press, 1997.

McDonnell, Janet. "World War II: Defending Park Values and Resources." *The Public Historian* 29, no. 4 (Fall 2007): 15–33.

Mendoza, Mary E. "Review of *City of Inmates: Conquest, Rebellion, and the Rise of Human Caging in Los Angeles, 1771–1965*, by Kelly Lytle Hernandez." *Environmental History* 24, no. 2 (April 2019): 397–99.

Meringolo, Denise. *Museums, Monuments, and National Parks: Toward a New Genealogy of Public History*. Amherst: University of Massachusetts Press, 2012.

Meyer, Jeffrey. "Alpine Experiments: The National Parks and the Development of Skiing in the American West." Master's thesis, University of Montana, Missoula MT, 2015.

Meyerson, Harvey. *Nature's Army: When Soldiers Fought for Yosemite*. Lawrence: University Press of Kansas, 2001.

Miles, John. *Wilderness in the National Parks: Playground or Preserve*. Seattle: University of Washington Press, 2009.

Miller, Warren. *Wine, Women, Warren, & Skis*. Vail, Colorado: Goldfinkle, Masowitch, Kloppenboig, & O'Brian, 1958.

Moses, L. C. *Wild West Shows and the Images of American Indians, 1883–1933*. Albuquerque: University of New Mexico Press, 1996.

Moss, Michael, and David Thomas, eds. *Archival Silences: Missing, Lost, and Uncreated Archives*. New York: Routledge, 2021.

Moss, Michael, and David Thomas. "Archival Silences and the National Parks." *Archives and Manuscripts* 42, no. 2 (2014): 56–76.

Muir, John. *The Mountains of California*. New York: Century Company, 1911.

———. *Our National Parks*. Boston: Houghton Mifflin and Company, 1901.

———. *The Yosemite*. New York: The Century Company, 1912. In *The Eight Wilderness Discovery Books*, by John Muir. Seattle: The Mountaineers, 1992.

National Park Service. *A Sense of Place: Design Guidelines for Yosemite Valley*. National Park Service, Yosemite National Park, 2012.

Nelson, Clair E. "Remembering the CCC: Buck Meadows, California, 1933–1934." *Journal of Forest History* 26, no. 4 (October 1982): 184–91.

O'Connell, Nicholas. *Beyond Risk: Conversations with Climbers*. Seattle: Mountaineers, 1993.

Ogden, Kate Nearpass. *Yosemite*. London: Reaktion Books Ltd., 2015.

Olmsted, Frederick Law. *The Papers of Frederick Law Olmsted*. Volume 5. Edited by Charles Capen McLaughlin, Charles E. Beveridge, David Schuyler, and Jane Turner Censer. Baltimore MD: Johns Hopkins University Press, 1977.

———. "Yosemite and the Mariposa Grove." In *Olmsted and Yosemite: Civil War, Abolition, and the National Park Idea*, edited by Rolf Diamant and Ethan Carr. Amherst: Library of American Landscape History, 2022.

Orr, Lois. "The Civilian Conservation Corps (CCC) in Yosemite." *Yosemite: A Journal for Members of the Yosemite Association* 67, no. 4 (Fall 2005): 3–7.

Orsi, Richard. *Sunset Limited: The Southern Pacific Railroad and the Development of the American West, 1850-1930*. Berkeley: University of California Press, 2005.

Ostler, Jeffrey. *Surviving Genocide: Native Nations and the United States from the American Revolution to Bleeding Kansas*. New Haven: Yale University Press, 2020.

Phillips, Katrina. *Staging Indigeneity: Salvage Tourism and the Performance of Native American History*. Chapel Hill: University of North Carolina Press, 2021.

Pyne, Stephen J. *Pyrocene Park: A Journey into the Fire History of Yosemite National Park*. Tucson: University of Arizona Press, 2023.

Report of the Commissioners to Manage the Yosemite Valley and the Mariposa Big Trees, 1889–90. Sacramento: Superintendent of State Documents, 1890.

Righter, Robert. *The Battle over Hetch Hetchy: America's Most Controversial Dam and the Birth of Modern Environmentalism*. New York: Oxford University Press, 2005.

Robbins, Roy M. *Our Landed Heritage: The Public Domain, 1776–1970*. Lincoln: University of Nebraska Press, 1942.

Roberts, Brian. *American Alchemy: The California Gold Rush and Middle-Class Culture*. Chapel Hill: University of North Carolina, 2000.

Roper, Laura Wood. *FLO: A Biography of Frederick Law Olmsted*. Baltimore: Johns Hopkins University Press, 1973.

Roper, Steve. *Camp 4: Recollections of a Yosemite Rock Climber*. Seattle: Mountaineers, 1994.

Rose, Gene. *Magic Yosemite Winters: A Century of Winter Sports* Truckee CA: Coldstream Press, 1999.

Rosenzweig, Roy, and Elizabeth Blackmar. *The Park and the People: A History of Central Park*. Ithaca NY: Cornell University Press, 1992.

Rosenzweig, Roy, and David Thelen. *The Presence of the Past: Popular Uses of History in American Life*. New York: Columbia University Press, 1998.

Rothman, Hal. *Blazing Heritage: A History of Wildland Fire in the National Parks*. Oxford: Oxford University Press, 2007.

———. *Devil's Bargains: Tourism in the Twentieth Century American West*. Lawrence: University Press of Kansas, 1998.

———. *Saving the Planet: The American Response to the Environment in the Twentieth Century*. Chicago: Ivan R. Dee, 2000.

Rugh, Susan Sessions. *Are We There Yet?: The Golden Age of American Family Vacations*. Lawrence: University Press of Kansas, 2008.

Runte, Alfred. *National Parks: The American Experience*. Lincoln: University of Nebraska Press, 1979.

———. *Yosemite: The Embattled Wilderness*. Lincoln: University of Nebraska Press, 1990.

Russell, Carl Parcher. *One Hundred Years in Yosemite: The Story of a Great Park and Its Friends*. Berkeley: University of California Press, 1947.

Sanborn, Margaret. *Yosemite: Its Discovery, Its Wonders and Its People*. Yosemite National Park: Yosemite Association, 1989.

Sargent, Shirley. *Galen Clark, Yosemite Guardian*. San Francisco: Sierra Club, 1964.

———. *Wawona Yesterdays* Yosemite National History Association, 1961.

———. *Yosemite's Innkeepers: The Story of a Great Park and Its Chief Concessionaires*. Yosemite CA: Ponderosa Press, 2000.

Sax, Joseph. *Mountains without Handrails: Reflections on the National Parks*. Ann Arbor: University of Michigan Press, 1980, 2018.

Scott, Amy. *Yosemite: Art of an American Icon*. Berkeley: University of California Press, 2006.

Scott, David, and Kang Jae Jerry Lee. "People of Color and Their Constraints to National Park Visitation." *George Wright Forum* 35, no. 1 (January 2018): 73–82.

Sears, John. *Sacred Places: American Tourist Attractions in the Nineteenth Century*. New York: Oxford University Press, 1989.

Sellars, Richard West. *Preserving Nature in the National Parks: A History*. New Haven: Yale University Press, 1997.

Shaffer, Marguerite. "Performing Bears and Packaged Wilderness: Reframing the History of National Parks." In *Cities and Nature in the American West*, edited by Char Miller. Reno: University of Nevada Press, 2010.

Shaffer, Marguerite S. *See America First: Tourism and National Identity, 1880–1940*. Washington DC: Smithsonian Institution Press, 2001.

Shankland, Robert. *Steve Mather of the National Parks*. 3rd ed. New York: Knopf Publishers, 1970.

Simpson, John. *Dam!: Water, Power, Politics, and Preservation in Hetch Hetchy and Yosemite National Park*. New York: Pantheon Books, 2005.

Skillen, James. *This Land is My Land: Rebellion in the West*. Oxford: Oxford University Press, 2020.

Smith, Conrad. *Media and Apocalypse: News Coverage of the Yellowstone Forest Fires, Exxon Valdez Oil Spill, and Loma Prieta Earthquake*. Contributions to the Study of Mass Media and Communications. Westport CT: Greenwood Publishing Group, 1992.

Solnit, Rebecca. *Savage Dreams: A Journey into the Landscape Wars of the American West*. Berkeley: University of California Press, 1999.

———. *Storming the Gates of Paradise: Landscapes for Politics*. Berkeley: University of California Press, 2007.

Soukup, Michael, and Gary E. Machlis. *American Covenant: National Parks, Their Promise, and Our Nation's Future*. New Haven: Yale University Press, 2021.

Spaulding, Jonathan. *Ansel Adams, and the American Landscape: A Biography*. Berkeley: University of California Press, 1995.

———. "Yosemite and Ansel Adams: Art, Commerce, and Western Tourism." *Pacific Historical Review* 65, no. 4 (November 1996): 615–39.

Spence, Mark David. *Dispossessing the Wilderness: Indian Removal and the Making of National Parks*. Oxford: Oxford University Press, 1999.

———. *Watershed Park: Administrative History of Redwood National and State Parks.* National Park Service, U.S. Department of the Interior, History Program, Pacific West Region, 2011.

Starr, Kevin. *Americans and the California Dream, 1850–1915*. New York: Oxford University Press, 1973.

State of California. *Biennial Report of the Commissioners to Manage the Yosemite Valley and the Mariposa Big Tree Grove for the Years 1874 and 1875*. Sacramento: G. H. Springer, State Printer, 1875.

Stegner, Wallace. *Uneasy Chair: A Biography of Bernard DeVoto*. Lincoln: University of Nebraska Press, 1973.

Steinberg, Ted. *Down to Earth: Nature's Role in American History*. New York: Oxford University Press, 2013.

Stine, Scott. *A Way across the Mountain: Joseph Walker's 1833 Trans-Sierran Passage and the Myth of Yosemite's Discovery*. Volume 25 in American Trails Series. Norman: University of Oklahoma Press, 2015.

Stoll, Mark. *Inherent the Holy Mountain: Religion and the Rise of American Environmentalism*. New York: Oxford University Press, 2015.

———. "Milton in Yosemite: *Paradise Lost* and the National Parks Idea." *Environmental History* 13, no. 3 (April 2008): 237–74.

Swan, Christopher, and Chet Roaman. *YV88: An Eco-Fiction of Tomorrow*. San Francisco: Sierra Club Books, 1977.

Switzer, Jacqueline Vaughn. *Green Backlash: The History and Politics of the Environmental Opposition in the U.S.* Boulder: Lynne Rienner Publishers, 1997.

Szarkowski, John. "Introduction Ansel Adams, *The Portfolios of Ansel Adams*." Boston: Little, Brown and Company, 1977.

Taylor, Joseph. *Pilgrims of the Vertical: Yosemite Rock Climbers and Nature at Risk*. Cambridge MA: Harvard University Press, 2010.

Thomas, David, Simon Fowler, and Valerie Johnson. *The Silence of the Archive*. London: Facet Publishing, 2017.

Tresidder, Donald. "The National Parks: A Public Health Problem." PhD diss., Stanford University, 1927.

Trexler, Keith A. *The Tioga Road: A History 1883–1961*. Yosemite Natural History Association, 1961.

Trouilot, Michel-Rolph. *Silencing the Past: Power and the Production of History*. Boston: Beacon Press 1995.

Tuan, Yi-Fu. "Place: An Experiential Perspective." *Geographical Review* 65, no. 2 (April 1975): 151–65.

Turner, Tom. *David Brower: The Making of the Environmental Movement*. Berkeley: University of California Press, 2015.

Tweed, William. *Uncertain Path: A Search for the Future of National Parks*. Berkeley: University of California Press, 2010.

Urry, John. *The Tourist Gaze 3.0*. London: Sage Publishing, 2011.

Watt, Laura Alice. *The Paradox of Preservation: Wilderness and Working Landscapes at Point Reyes National Seashore*. Berkeley: University of California Press, 2016.

Watts, Jennifer. "Photography's Workshop: Yosemite in the Modern Era." In *Yosemite: Art of an American Icon*, edited by Amy Scott. Berkeley: University of California Press, 2006.

West Sellers, Richard. *Preserving Nature in the National Parks: A History*. New Haven: Yale University Press, 1987.

Whitney, J. D. *Report of the Commissioners to Manage the Yosemite Valley and the Mariposa Big Tree Grove*. Sacramento: D. W. Gelwicks, State Printer, 1867.

Whitney, Josiah. *The Yosemite Book: A Description of the Yosemite Valley and the Adjacent Region of the Sierra Nevada, and the of the Big Trees of California*. New York: Julius Bien, 1868.

Wilkins, Emily, Jordan W. Smith, and Rose Keane. "Social Media Communication Preferences of National Park Visitors." *Applied Environmental Education and Communication* 19, no. 1 (2020): 4–18.

Wirth, Conrad L. *Parks, Politics, and the People*. Norman: University of Oklahoma Press, 1980.

Wolfe, Linnie Marsh. *Son of the Wilderness: The Life of John Muir*. New York: A. A. Knopf, 1945.

Worster, Donald. *A Passion for Nature: The Life of John Muir*. New York: Oxford University Press, 2008.

Wrenn, Rebecca Louise. "Playing in the Park: Winter Sports and Sport Spectacles in Yosemite, 1900–1950." PhD diss., University of California, Riverside, 2017.

Wrobel, David M. *Global West, American Frontier: Travel, Empire, and Exceptionalism from Manifest Destiny to the Great Depression*. Calvin P. Horn Lectures in Western History and Culture. Albuquerque: University of New Mexico Press, 2013.

Wyss, Robert. *The Man Who Built the Sierra Club: A Life of David Brower*. New York: Columbia University Press, 2016.

Young, Terrence. *Heading Out: A History of American Camping*. Ithaca: Cornell University Press, 2017.

Index

Illustrations are indicated by F *with a numeral; all appear following pages 70 and 130*

In the Environment and Region in the American West series

Losing Eden: An Environmental History of the American West, New Edition
Sara Dant
Foreword by Tom S. Udall

The Mountains Are Calling: Tourists and the Unmaking of Yosemite National Park
Michael W. Childers

www.ingramcontent.com/pod-product-compliance
Lightning Source LLC
LaVergne TN
LVHW091514200926
848483LV00008B/458/J
* 9 7 8 1 4 9 6 2 3 9 5 8 7 *